New Generation Coatings for Metals

New Generation Coatings for Metals

Editors

Anthony E. Hughes
Russel Varley

MDPI • Basel • Beijing • Wuhan • Barcelona • Belgrade • Manchester • Tokyo • Cluj • Tianjin

Editors
Anthony E. Hughes
Institute for Frontier Materials,
Deakin University
Australia

Russel Varley
Institute for Frontier Materials,
Geelong Waurn Ponds Campus,
Deakin University
Australia

Editorial Office
MDPI
St. Alban-Anlage 66
4052 Basel, Switzerland

This is a reprint of articles from the Special Issue published online in the open access journal *Coatings* (ISSN 2079-6412) (available at: https://www.mdpi.com/journal/coatings/special_issues/coatings_metals).

For citation purposes, cite each article independently as indicated on the article page online and as indicated below:

LastName, A.A.; LastName, B.B.; LastName, C.C. Article Title. *Journal Name* **Year**, *Article Number*, Page Range.

ISBN 978-3-03943-274-5 (Hbk)
ISBN 978-3-03943-275-2 (PDF)

© 2020 by the authors. Articles in this book are Open Access and distributed under the Creative Commons Attribution (CC BY) license, which allows users to download, copy and build upon published articles, as long as the author and publisher are properly credited, which ensures maximum dissemination and a wider impact of our publications.

The book as a whole is distributed by MDPI under the terms and conditions of the Creative Commons license CC BY-NC-ND.

Contents

About the Editors vii

Preface to "New Generation Coatings for Metals" ix

Yingying Wang, Bernard Normand, Xinkun Suo, Marie-Pierre Planche, Hanlin Liao and Junlei Tang
Cold-Sprayed AZ91D Coating and SiC/AZ91D Composite Coatings
Reprinted from: *Coatings* **2018**, *8*, 122, doi:10.3390/coatings8040122 1

Liang Hao, Hiroyuki Yoshida, Takaomi Itoi and Yun Lu
Preparation of Metal Coatings on Steel Balls Using Mechanical Coating Technique and Its Process Analysis
Reprinted from: *Coatings* **2017**, *7*, 53, doi:10.3390/coatings7040053 15

Martin Buchtík, Petr Kosár, Jaromír Wasserbauer, Jakub Tkacz and Pavel Doležal
Characterization of Electroless Ni–P Coating Prepared on a Wrought ZE10 Magnesium Alloy
Reprinted from: *Coatings* **2018**, *8*, 96, doi:10.3390/coatings8030096 25

Yung-I Chen, Tso-Shen Lu and Zhi-Ting Zheng
Internally Oxidized Ru–Zr Multilayer Coatings
Reprinted from: *Coatings* **2017**, *7*, 46, doi:10.3390/coatings7040046 39

Domna Merachtsaki, Panagiotis Xidas, Panagiotis Giannakoudakis, Konstantinos Triantafyllidis and Panagiotis Spathis
Corrosion Protection of Steel by Epoxy-Organoclay Nanocomposite Coatings
Reprinted from: *Coatings* **2017**, *7*, 84, doi:10.3390/coatings7070084 51

Mona Taghavikish, Naba Kumar Dutta and Namita Roy Choudhury
Emerging Corrosion Inhibitors for Interfacial Coating
Reprinted from: *Coatings* **2017**, *7*, 217, doi:10.3390/coatings7120217 71

Xiaohui Liu, Shuaixing Wang, Nan Du, Xinyi Li and Qing Zhao
Evolution of the Three-Dimensional Structure and Growth Model of Plasma Electrolytic Oxidation Coatings on 1060 Aluminum Alloy
Reprinted from: *Coatings* **2018**, *8*, 105, doi:10.3390/coatings8030105 99

Anthony Hughes, James Laird, Chris Ryan, Peter Visser, Herman Terryn and Arjan Mol
Particle Characterisation and Depletion of Li_2CO_3 Inhibitor in a Polyurethane Coating
Reprinted from: *Coatings* **2017**, *7*, 106, doi:10.3390/coatings7070106 113

About the Editors

Anthony E. Hughes currently holds honorary positions of Adjunct Professor at Deakin University within the Institute for Frontier Materials, Deakin University and is a CSIRO Fellow at CSIRO Mineral Resources. He qualified with a Bachelor of Applied Science (Phys) with distinction from the Royal Melbourne Institute of Technology (RMIT) in 1978. In 1982, he obtained a Master in Applied Science, also from RMIT, with a thesis entitled *Photon Correlation Spectroscopy Study of Polymer Solutions*. He was the first Ph.D. graduate from the newly formed RMIT university when he was granted a Ph.D. for his thesis entitled *X-ray Photoelectron Spectroscopy Study of Segregation Phenomena in Yttria-Zirconia Solid Electrolytes* (1991). He was employed as an electron spectroscopist (X-ray Photoelectron Spectroscopy) at CSIRO Division of Materials Science in 1982, where he worked on characterization studies of catalyst and catalytic reactions from fundamental perspectives of single-crystal studies to more practical catalysts, where he examined the dispersion of catalytic metals on mostly oxide supports. In 1990, Professor Hughes began working in corrosion inhibitors, corrosion management systems, and coatings almost exclusively for aerospace applications. This work was undertaken with aircraft manufacturers such as Boeing (1990–1996) and BAE SYSTEM (1999–2003). It resulted in several patents for conversion coating processes and one commercial Ce-based process developed with Chemetall GmbH from 1997 to 1999. This process was eventually superseded by other processes where the bath chemistry was easier to maintain. The work with BAE SYSTEMS went a long way to developing prognostic health management systems for airframes. From 2007 to 2009, Professor Hughes was an external participant in a Dutch Program on Self Healing Materials working on a project managed by Professor J.M.C. Mol at TU Delft, called *Pre-emptive healing by responsive release in organic coatings*. This project delivered new, high-throughput techniques using multi-electrodes and microfluidics, both of which are used more broadly by other research groups, now in combination with neural networks as discovery methods. In the TU Delft/CSIRO project some 6000 combinations of inhibitors under various conditions were tested. Between 2010–2013, Professor Hughes was involved in a long-term commercial project on high throughput catalysis research for new ammonia synthesis approaches. In addition, he was also leading a large commercial research project looking to find new multifunctional inhibitors to replace chromate as inhibitors in aerospace primer paint coatings, as well as demonstrating self-healing mechanisms. From 2006 to his retirement in 2014, Professor Hughes moved into research management, where he was required to assemble diverse research areas into groups under a thematic management structure aimed at industry sectors. Such structures included *Interfacial Science, Corrosion Science and Surface Design, Nanoadditives for the Fine Chemical Industry*, and *Cleantech*. The largest of these groupings had over 55 staff grouped into 15 to 20 research areas. Since retiring in 2014, Professor Hughes has undertaken honorary positions at Deakin University (additive manufacturing, pipeline corrosion and inhibition) and CSIRO (aluminum alloy microstructure and corrosion). At Deakin University he supervises students working on advanced manufacturing of stainless steels using 3D laser printing. He contributes extensively to the 3D characterization of these materials using X-ray CT techniques. He has written around 200 papers and book chapters. He has eight patent families and is also one of only 3 recipients of the RSC Corrosion award (2005).

Russell J. Varley earned his B.Sc (Hons) in Physical and Inorganic Chemistry from The University of Adelaide, South Australia in 1987. After graduation, he worked for 1.5 years at Yorkshire Chemicals in Melbourne, Victoria, a manufacturer of chemicals for the textile industry, before gaining employment at the CSIRO Division of Chemicals and Polymers in 1989. While still working as a researcher at CSIRO he completed his Ph.D. in Materials Engineering in 1998 from Monash University, Australia entitled *Thermoplastic Toughening of a Tri-functional Epoxy Resin*. His research career began by working on the development of new polymer matrices for high-temperature composites for aerospace applications, which led to a life-long interest in using polymer chemistry to design processable, high-performance, and functional polymer composites. After spending several months at the Technical University of Delft, the Netherlands in 2007 working with Prof Sybrand van der Zwaag he became passionate about applying self-healing methods to create polymer composites and coatings for demanding applications. In 2016, he left his position as the leader of the Industrial Composites team at CSIRO to join the Institute for Frontier Materials, Deakin University as Professor of Composite Materials. Since then, his research goals have expanded to include the synthesis of next-generation carbon fiber using novel rapid oxidation methods, alternative precursors, and novel architectures for enhanced sustainability and improved structural and functional properties. He has published more than 100 papers in peer-reviewed literature, 3 book chapters, and 3 patents. As Carbon Nexus group leader he leads a portfolio of externally funded projects focused on industrially relevant outcomes within the aerospace, automotive, renewable energy, and oil and gas sectors of the economy. In 2009, he won the CSIRO Medal for Business Excellence as part of the PETRONAS Engagement team and, in 2013, won the CSIRO Newton Turner Award for Scientific Excellence.

Preface to "New Generation Coatings for Metals"

The world of coatings research is enormous. Coating applications span diverse areas such as nanolayers on single-crystal surfaces to organic coatings for the protection of offshore structures, such as wind turbines and drilling platforms, to heat resistant and self-healing coatings and materials for space applications. The purpose of coatings can probably be divided into two groups, cosmetic and functional. In the former case, the coating simply adds to the aesthetic of the substrate; colored paint for a wall in a house is a good example of a cosmetic coating. Functional coatings serve a purpose other than cosmetic. They can broadly be defined as extending the lifetime of the underlying substrate in a particular environment and under certain stressors. Such examples might include wear-resistant coatings applied to bearings, heat-resistant coatings applied to turbine blades to reduce thermal shock and oxidation in jet engines, corrosion-resistant coatings that prevent metal corrosion, and localized attack on metals. It may also include camouflage coatings where the coating itself disguises the substrate to minimize the risk of physical attack. Stealth coatings would fall into this category. Given the wide range of applications, it is not surprising to discover that there is also a wide range of substrates that require protection through the application of coatings including bio-derived materials, polymers and metals.

The coatings themselves draw from most scientific disciplines such as physics, chemistry, biology, and materials science, adapting many principles from these disciplines for problem-solving in coatings. This brings us to this short book. The reader will not be surprised, given the broad nature of coatings, that the research presented in this book is also of a broad nature. To begin, we look at two papers that investigate metallic coatings on different substrates. Yingying Wang et al. examine AZ91D (a magnesium alloy containing Al and Zn) coatings on steel and magnesium substrates. These coatings are deposited using cold spray technology where the AZ91D powder is accelerated to the target material as high velocity. Cold spray is an emerging technique that shows promise for coating manufacture as well as repair. In the second contribution, Liang Hao et al. use a mechanical plating method to apply Ti-based coatings onto steel balls as part of a wider program of understanding how mechanical plating works. The third paper, by Buchtik et al., explores how improved chemical and physical resistance can be obtained by applying an electroless Ni-P coating to a magnesium substrate. Electroless plating is a chemical process performed in a coating bath. The fourth contribution by Yung-I Chen et al. uses the coatings technique of magnetron sputtering. In this paper, the authors study the structure and hardness of multilayer structures consisting of a 100 nm Cr base layer and a Ru-Zr overlayer.

The next contribution by Merachtsaki et al. examines the corrosion protection of steel using a nanocomposite comprising epoxy and clay nanoparticles modified by organo-ammonium ions. The following paper is an excellent review paper by Taghavikish et al. on ionic and polyionic liquids for corrosion protection. The penultimate paper, by Xiaohui Liu et al., looks at the growth models of plasma electrolytic oxidation (PEO) coatings grown on an aluminum alloy. PEO is another relatively new coating method that promotes oxide formation during the electrical breakdown in an electrolyte. The final paper, by the editor and co-authors, looks at the characterization of the leaching of a new, non-chromate, Li-based inhibitor from a traditional polyurethane-based primer for aerospace applications. In summary, the papers presented in this volume will provide the reader with detailed studies of the breadth and type of work that is being undertaken using a variety of approaches in the field of coatings.

Anthony E. Hughes, Russel Varley
Editors

Article

Cold-Sprayed AZ91D Coating and SiC/AZ91D Composite Coatings

Yingying Wang [1], Bernard Normand [2], Xinkun Suo [3,*], Marie-Pierre Planche [4], Hanlin Liao [4] and Junlei Tang [1,*]

1. College of Chemistry and Chemical Engineering, Southwest Petroleum University, Chengdu 610500, China; yingyingwanglyon@126.com
2. Université de Lyon, INSA de Lyon, MATEIS CNRS UMR 5510, 69621 Villeurbanne, France; bernard.normand@insa-lyon.fr
3. Key Laboratory of Marine Materials and Related Technologies, Key Laboratory of Marine Materials and Protective Technologies of Zhejiang Province, Ningbo Institute of Materials Technology and Engineering, Chinese Academy of Sciences, Ningbo 315201, China
4. ICB UMR 6303, CNRS, University Bourgogne Franche-Comté, UTBM, 90100 Belfort CEDEX, France; marie-pierre.planche@utbm.fr (M.-P.P.); hanlin.liao@utbm.fr (H.L.)
* Correspondence: suoxinkun@gmail.com (X.S.); tangjunlei@126.com (J.T.); Tel.: +86-158-2557-3811 (X.S.); +86-186-0803-9391 (J.T.)

Received: 21 January 2018; Accepted: 23 March 2018; Published: 26 March 2018

Abstract: As an emerging coating building technique, cold spraying has many advantages to elaborate Mg alloy workpieces. In this study, AZ91D coatings and AZ91D-based composite coatings were deposited using cold spraying. Coatings were prepared using different gas temperatures to obtain the available main gas temperature. Compressed air was used as the accelerating gas, and although magnesium alloy is oxidation-sensitive, AZ91D coatings with good performance were obtained. The results show that dense coatings can be fabricated until the gas temperature is higher than 500 °C. The deposition efficiency increases greatly with the gas temperature, but it is lower than 10% for all coating specimens. To analyze the effects of compressed air on AZ91D powder particles and the effects of gas temperature on coatings, the phase composition, porosity, cross-sectional microstructure, and microhardness of coatings were characterized. X-ray diffraction and oxygen content analysis clarified that no phase transformation or oxidation occurred on AZ91D powder particles during cold spraying processes with compressed air. The porosity of AZ91D coatings remained between 3.6% and 3.9%. Impact melting was found on deformed AZ91D particles when the gas temperature increased to 550 °C. As-sprayed coatings exhibit much higher microhardness than as-casted bulk magnesium, demonstrating the dense structure of cold-sprayed coatings. To study the effects of ceramic particles on cold-sprayed AZ91D coatings, 15 vol % SiC powder particles were added into the feedstock powder. Lower SiC content in the coating than in the feedstock powder means that the deposition efficiency of the SiC powder particles is lower than the deposition efficiency of AZ91D particles. The addition of SiC particles reduces the porosity and increases the microhardness of cold-sprayed AZ91D coatings. The corrosion behavior of AZ91D coating and SiC reinforced AZ91D composite coating were examined. The SiC-reinforced AZ91D composite coating reveals higher corrosion potential than magnesium substrate; therefore, it serves as a cathode for the magnesium substrate, the same as the AZ91D coating on magnesium substrate. As the SiC powder is semi-conductive, the embedded SiC particles reduce the electrochemical reaction of the AZ91D coating. The addition of SiC particles increases the corrosion potential of the coating, meanwhile increasing the galvanic potential and decreasing the negative galvanic current of the coating-substrate couple.

Keywords: cold spraying; coating; magnesium alloy; composite coatings; corrosion; microstructure

1. Introduction

Magnesium (Mg) alloys have potential use in aircraft and automobile industries due to their excellent specific strength [1]. They are also used to protect steel structures from corrosion in freshwater and soil environments as sacrificial anodes [2]. Casting is the main technique to produce AZ91D alloy workpieces; however, it is not suitable to fabricate large or complex-shaped products [3]. In addition, AZ91D work pieces are difficult to be repaired using traditional techniques (tungsten inert gas (TIG) welding [4], metal inert gas (MIG) welding [5], and thermal spraying [6], etc.) due to their high-temperature activity.

As an emerging coating building and rebuilding technique, cold spraying has advantages to elaborate Mg alloy workpieces [7,8]. In cold spraying, particles are accelerated to a high velocity (300~1200 m·s^{-1}) by a high-pressure thermal gas through a de Laval-type nozzle. Then, the particles can deposit onto a substrate and form a coating once the particle velocity exceeds the critical velocity of the material [9,10]. In the view of bonding, the jet formation of impact couples is viewed as the necessary condition.

In cold spraying, the main gas temperature is lower than the melting point of the sprayed material; thus, oxidization can be avoided. In the past decade, cold spraying has been used to deposit oxidation-sensitive alloys, such as Ti [11], Al [12,13], and Mg [8] alloy coatings. In addition, cold spraying was successfully used to repair Al alloy gearbox components [14]. However, few publications on cold-sprayed AZ91D coatings and AZ91D-based composite coatings were reported. Based on this fact, AZ91D powder was used to elaborate coatings using cold spraying in this paper, and SiC powder was used as reinforcing particles to deposit the composite coatings. In consideration of the application of cold-sprayed Mg alloy workpieces, compressed air, not noble gas, was used as the accelerating gas. Different gas temperatures were conducted to study the feasibility of fabricating cold-sprayed Mg alloy work pieces, and then ceramic particles were added as reinforcement to improve the performance of the coatings. The effects of gas temperature and the addition of ceramic particles on the coatings' microstructure and electrochemical behaviors were evaluated.

2. Materials and Methods

2.1. Materials

A commercially-available AZ91D powder (WeiHao Magnesium Powder Co., Ltd., Tangshan, China) and SiC powder (Sulzer Metco, Bron, France) were used to produce coatings. The AZ91D particles presented a spherical shape (Figure 1a), and the grain orientation is random (Figure 1b). SiC particles show dendritic morphology with sharp edges (Figure 1c). Size distribution of the powder was examined using a MASTERSIZER 2000 system (Malvern Panalytical Ltd., Royston, UK). The size distribution of the AZ91D powder was in the range from 34 µm to 81 µm, and averaged 52 µm (Figure 1d). The size distribution of SiC powder is from of 4 µm to 55 µm, and averaged 15 µm (Figure 1e). The content of SiC particles in the feedstock powder was 30 vol %. Stainless steel and magnesium plates with a dimensions of 60 mm × 20 mm × 2 mm were used as substrates. The substrates were sandblasted using alumina grits (ISO 6344, Grit designation, P100) prior to spraying.

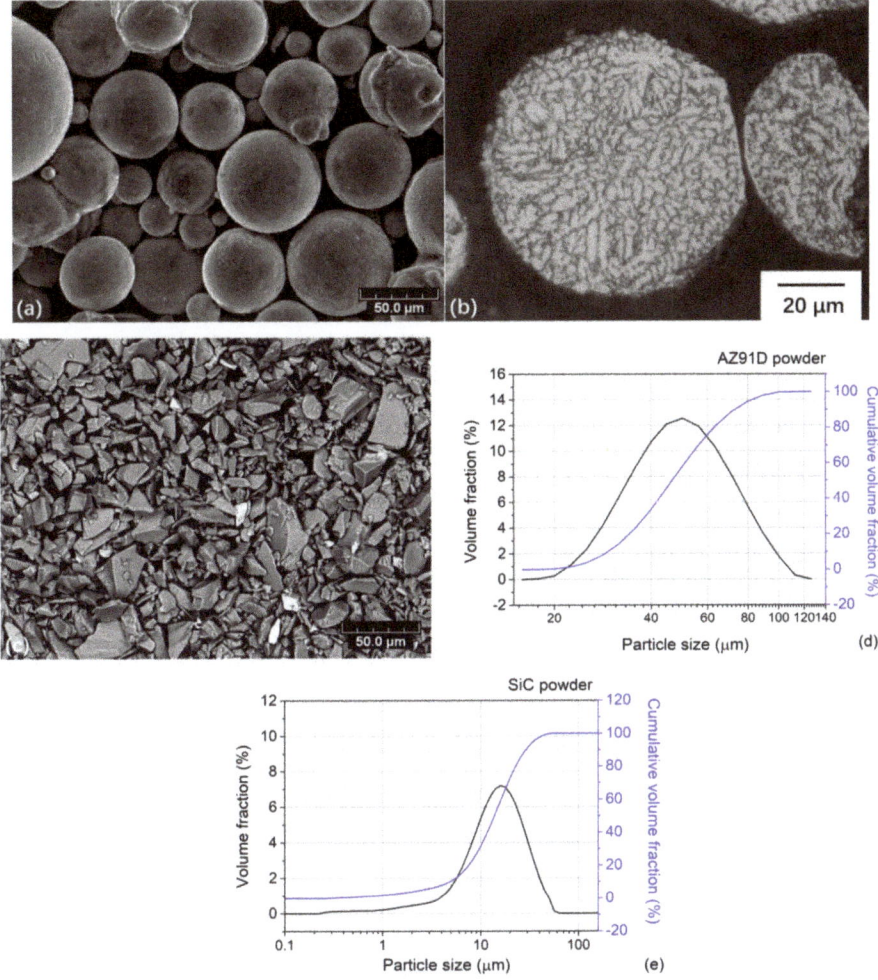

Figure 1. Morphology of the original AZ91D powder (**a**); sectional microstructure of an AZ91D particle after etching (**b**); and the morphology of SiC powder (**c**). Size distribution of the AZ91D powder particles (**d**) and SiC powder particles (**e**).

2.2. Spraying Conditions

A self-designed rectangular cross-section nozzle was used to fabricate the coatings. The expansion ratio of the nozzle is 4.9. The outlet diameter of the nozzle is 6.5 mm and the divergent section length is 170 mm. Compressed air and argon were used as the main gas and the carrier gas, respectively. The main gas pressure in a pre-chamber was 2.5 MPa, and the gas temperatures were changed in a range from 450 °C to 600 °C. The powder feed rate was 70 g·min^{-1}. The standoff distance was 30 mm. The nozzle was fixed on a robot in order to maintain a traverse speed of 100 mm·s^{-1}. The beam distance was 2 mm. Each coating contains two layers.

2.3. Coating Characterization

The particle morphology and coating fracture morphologies were observed using a scanning electron microscope (SEM) (JEOL, JSM-5800LV, Tokyo, Japan). The coatings cross-sections were observed using an optical microscope (OM) (Nikon, Tokyo, Japan) and SEM. Porosity was estimated using an open software, Image J (National Institutes of Health, Bethesda, MD, USA). X-ray diffraction (XRD, D/mas-2400, Rigaku, Tokyo, Japan) was used to determine the phases in the feedstock powder and coatings. Oxygen contents in both the feedstock powder and cold-sprayed coatings were measured using an oxygen, nitrogen, and hydrogen analyzer (ONH-2000, ELTRA, Haan, Germany). The microhardness of the coatings was measured under a load of 300 g with a dwell time of 15 s. For each coating, three specimens were used to measure the microhardness. In addition, for each specimen, 10 measurements were conducted in random locations to obtain the average microhardness.

To evaluate the corrosion protective effects of the Mg alloy coatings, electrochemical measurements were performed by a standard three-electrode electrochemical cell in air saturated with a 0.1 M Na_2SO_4 solution. Coating specimens were conducted as the working electrode with a surface of 10 mm × 10 mm. The counter electrode was a platinum sheet with a surface of 8 mm × 8 mm. The reference electrode was a saturated mercury sulfate electrode (Hg/Hg_2SO_4). The solution concentration inside the reference electrode compartment was a saturated K_2SO_4 solution, with a potential of 640 mV with respect to the normal hydrogen electrode (NHE) and a potential of 400 mV with respect to the saturated calomel electrode (SCE). The open circuit potential (OCP) and polarization scans were conducted on an electrochemical workstation (Reference 600TM, Potentiostat/Galvanostat/ZRA, Gamry Instruments, Inc., Warminster, PA, USA). Potentiodynamic scans were performed from −0.1 V to 2.5 V (vs. OCP) with a scan rate of 1 mV/s. Galvanic interaction between the coatings and the substrate has been studied by continuous monitoring of the galvanic potential and the galvanic current using ZRA measurements.

3. Results

3.1. Effects of Gas Temperature on the AZ91D Coatings

The microstructures of the coatings elaborated at 500 °C, 550 °C, and 600 °C are shown in Figure 2a–c, respectively. It was found that the coating could be built on stainless steel until the main gas temperature increased to 500 °C. It is reported that in cold spraying the particle velocity increases as the main gas temperature increases. Thus, it can be deduced that the particle velocity was lower than the critical velocity of magnesium when the main gas temperature was lower than 500 °C. The critical velocity of Mg was reported in a range of 653 m/s to 677 m/s, which was calculated by simulation [8]. The particle velocity increased higher than the critical velocity as main gas temperature is higher than 500 °C; thus, coatings could be built.

In Figure 2, the part between the dark region on the top and the uniform gray area at the bottom is the coating. The thickness increases with the gas temperature as the average coating thickness is 177.2 ± 30.7, 482.8 ± 69.5, and 516.5 ± 50.4 μm when the gas temperature is 500 °C, 550 °C, and 600 °C. No visible micro–channels in the entire view of section can be observed for all coating specimens. At the coating/substrate interface, there are neither cracks nor delamination. The coatings reveal a dense microstructure. Several dark dots are identified as pores. The AZ91D powder particles are fully deformed as the original spherical shape can no longer be recognized. The gains of the AZ91D powder particles are elongated along the direction vertical to their impacting direction.

Figure 3 shows the analysis of the grain in a projected and bounced particle on the substrate. The white arrow indicates the direction of impact on the particle. It was found that not only the shape of the particles changes, but also the orientation of the grains. Within the original particle, the grain orientation is random. After impact, an ordered arrangement in the impact zone is presented and the grain arrangement direction is perpendicular to the direction of the rebound. As a result, AZ91D powder particles can deform and form a cold-sprayed coating although it is difficult to deform.

Figure 2. Microstructures of coatings deposited at 500 °C (**a**), 550 °C (**b**), and 600 °C (**c**).

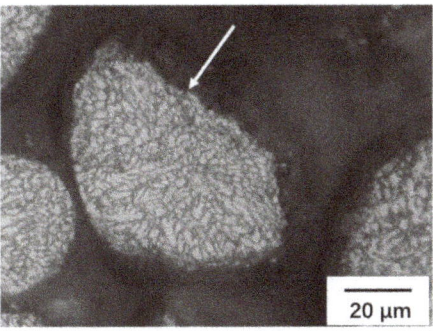

Figure 3. Morphology of a particle after the rebound.

The coating porosities were calculated. The average porosities were 3.9% (±0.7%), 3.6% (0.6%), and 3.7% (0.1%) when the gas temperature was 500 °C, 550 °C, and 600 °C, respectively. It is found that coating porosity decreased with an increase in the main gas temperature. In this study, main gas temperatures were restricted in a narrow range: on one hand, it must ensure particle deposition; on the other hand, it should be lower than the melting point of the AZ91D particles. The narrow temperature interval resulted in a narrow particle velocity range which, in turn, resulted in the slight variation of the porosities. Additionally, the porosity of the AZ91D coating is close to the porosity of the pure magnesium coating [7]. When compared with another widely-used sacrificial coating, i.e., aluminum alloy (the porosity is 1~2%), it is slightly larger [12,15].

The effects of gas temperature on the deposition efficiency of cold-sprayed AZ91D coatings are presented in Figure 4. At 450 °C, no coatings were obtained. From 500–600 °C, deposition efficiency increases with gas temperature. The deposition efficiency is low, which is lower than 10% for all specimens at different gas temperature. This means the AZ91D alloy is difficult to deposit using the cold spraying conditions in this study. Thanks to the recyclability of feedstock powder in cold spraying, we consider that the deposition efficiency would not restrict the application of cold-sprayed Mg alloy workpieces. Moreover, we have obtained high deposition efficiency for pure magnesium, whose deposition efficiency is approximately 60% at 600 °C [7]. The difference in deposition efficiency could be due to the different original shape of powder particles. In [7], the pure magnesium powder particles show an irregular shape instead of a spherical shape. It is reasonable to envisage the possibility of a higher deposition efficiency. When compared with the other materials in the literature, because of the weaker plastic deformation ability, the deposition efficiency is lower than that of cold-sprayed aluminum coatings deposited under similar conditions [16]. The gas temperature is 400 °C and 500 °C, and the gas pressure is 2.1–2.9 MPa in [16]. The deposition efficiency is even lower than that of cold-sprayed titanium alloy [12], however, the gas pressure is much higher, which is 4 MPa, as reported.

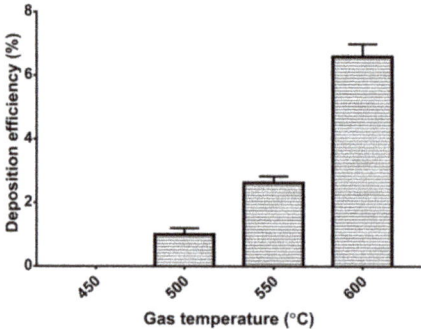

Figure 4. Deposition efficiency of AZ91D coatings deposited at different gas temperatures.

XRD and oxygen content analysis were employed to evaluate the phase transformation and oxidization of the particles during cold spraying. XRD patterns of the powder particles and coatings are shown in Figure 5. It can be found that there were only Mg and $Mg_{17}Al_{12}$ phases in the coatings, which were the same as those in the original particles. This allows the conclusion that there was no reaction or phase transformation during cold spraying [17,18]. Another interesting phenomenon was the peak intensity evolution of Mg before and after spraying. The peak intensity of crystal plane (0002) in the coatings obviously increased compared to that in the powder original. This may be due to the texture caused by the plastic deformation of the particles during impact [19]. Similar phenomena were also found in pure Mg coatings fabricated using cold spraying [8]. The texture depends on the deformation rate and velocity. In cold spraying, a typical particle/substrate contact time was 40 ns [20]. The intensive deformation happened during this time at a high velocity. Therefore, the deformation texture occurred.

The result of oxygen content analysis shows that the oxygen content in the coating produced at 600 °C (273 ± 55 ppm) was similar to the value in the original particles (200 ± 107 ppm). The in-flight time of Mg particles in the nozzle was less than 1 ms; additionally, the main gas temperature was lower than the melting point of the material. Therefore, the coatings did not undergo deterioration or oxidization. This is the same in cold-sprayed pure magnesium coating. The oxygen content of the Mg coating deposited at 630 °C is reported as 229 ± 124 ppm, which is much lower than the oxygen content of original Mg powder (866 ± 122 ppm). The decrease of oxygen content benefits from the fragmentation of the oxide film of the Mg particles. During cold spraying, the oxide film of the particles can be broken up, and metallurgical bonding could be achieved. Therefore, a higher particle velocity induced by a higher main gas temperature could result in a greater fragmentation of the oxide film. From the oxygen content results, the possibilities of the oxide film breaking up is much smaller for AZ91D powder particles in this study.

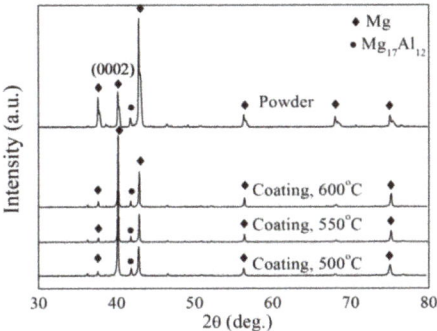

Figure 5. XRD patterns of AZ91D powder and AZ91D coatings deposited at different temperatures.

The fracture morphologies of the coatings elaborated at 500, 550, and 600 °C are shown in Figure 6. Deformed particles can be found everywhere in Figure 6a. No ductile fracture features can be found in this image. Therefore, the deposition of AZ91D coatings fabricated at 500 °C could be mainly due to the mechanical interlocking and jet mixing. An obvious laminar manner could be observed in the fracture surfaces. This is evidence of the plastic deformation.

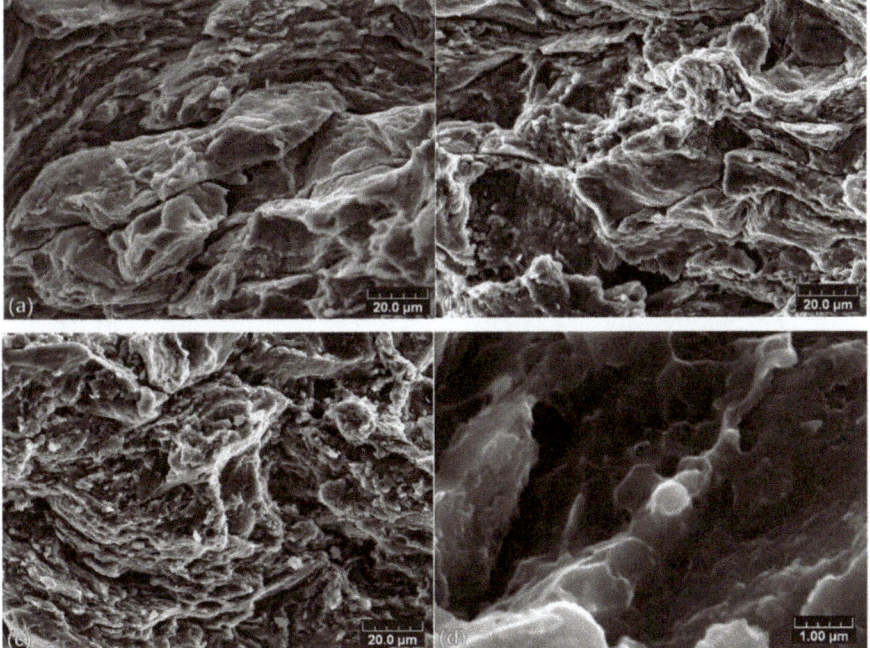

Figure 6. Fracture microstructures of AZ91D coatings elaborated at 500 °C (**a**), 550 °C (**b**), 600 °C (**c**); and partial enlarged drawing in **b** (**d**).

A partial enlarged image of Figure 6b is shown in Figure 6d. Dimple structures can be observed. The impact melting was found in many impact couples, especially in the impact of Al-based materials due to their low melting point. The melting point of AZ91D particles is lower than Al; thus, impact melting possibly occurred. According to the modeling result, the interfacial temperature of the impact couples could be close to the melting point of the impact materials [18]. The particle velocity increased as the main gas temperature increased to 550 °C; thus, some particles had more kinetic energy to achieve a higher deformation and, consequently, resulted in a higher interface temperature. This contributed to the partial impact melting between AZ91D particles. However, the dimple structures were very few. Therefore, the impact melting did not play a key role in bonding. The bonding mechanism of the coating is mainly the mechanical interlocking. It can be estimated that the bond of the coating and the substrate or between the particles is weak.

The microhardness of the AZ91D coatings elaborated at different temperatures was measured, and the result is shown in Figure 7. The microhardness of the coatings was about 100 HV, which was lower than the stainless steel substrate, but higher than the microhardness of as-casted bulks, which is 63.7 HV as reported by Masaki Sumida [21]. This is due to the work-hardening effect of subsequent particles during cold spraying. The same phenomenon was also found in cold-sprayed Al coating, Cu coating, and Fe coating [22].

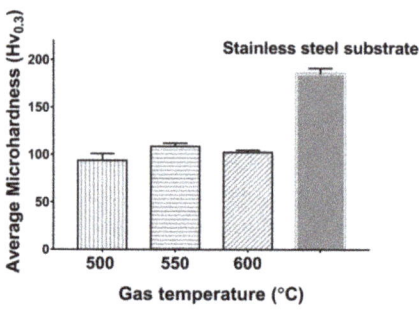

Figure 7. Microhardness of AZ91D coatings deposited at different temperatures.

Less dense coating with higher porosity results in lower microhardness. It can also be found that the microhardness of the coatings did not vary obviously as a function of the main gas temperature. This is due to the similar coating porosity in the narrow gas temperature range. Moreover, the microhardness of the AZ91D coating is much higher than the Mg coating using the same cold spraying conditions. The microhardness of the cold-sprayed Mg coating is approximately 38 HV [8].

3.2. Effects of Ceramic Particles Addition on AZ91D Coatings

AZ91D and SiC/AZ91D composite coatings were produced on a magnesium substrate to study the effects of ceramic particles addition. Similar to the AZ91D coating, the as-sprayed SiC/AZ91D composite coating exhibited a relatively rough outer surface. Figure 8a,b shows the top surface morphology of the AZ91D coating and SiC/AZ91D composite coating. The majority of particles on the top layer adhere well. Crevices could be detected beside loosely-adhered particles. The porous top layer is formed due to the absence of a tamping effect of the next incoming particles. SiC particles can be easily found on the top surface, and which are randomly distributed.

The entire thickness of SiC/AZ91D composite coating is presented in Figure 8c. The coating reveals a fully-dense microstructure. SiC particles and a small quantity of pores are distributed homogeneously within the coating. The AZ91D particles are fully deformed, and the original spherical shape cannot be identified any more. SiC particles keep the irregular morphology as before the cold spraying process. Due to the good adhesion between the coating and the substrate, the interface is difficult to detect.

Figure 8. Top surface morphology the AZ91D coating (**a**), and SiC/AZ91D composite coating (**b**); and the cross–sectional microstructure SiC/AZ91D composite coating (**c**).

The result of the image analysis shows that the content of SiC powder particles is 10.7 ± 3.2 vol %, which is lower than the SiC content in the feedstock blended powder. It is deduced that the deposition efficiency of the SiC powder particles is lower than the deposition efficiency of AZ91D particles. The same result was found in the literature [13,23] in which the SiC content in feedstock is 30 vol %, while in the composite coating, the SiC content is less than 30%.

The porosity and microhardness were measured on the SiC/AZ91D composite coating. The porosity of the composite coating is 1% (±0.14%), and the microhardness is 139 HV (±16.7 HV). The addition of SiC particles involve a significant coating densification and increasing microhardness. The coating porosity decreases with SiC addition during spraying. On one hand, the addition of SiC particles increases the deformation of the AZ91D powder particles, so that the porosity of the composite coatings becomes lower than that of the AZ91D coating due to the peening effects of the SiC particles. On the other hand, the different morphologies and sizes of the Al 5056 matrix powders and mixed SiC particles also play an important role. The microhardness of the composite coating increased greatly compared with that of the AZ91D coating, which should also be attributed to the reinforcing effect of SiC particles and work hardening effect of cold spraying. The uniformly distributed SiC particles could restrict the deformation of the AZ91D matrix. Additionally, the dislocation tangle resulting from the working-hard effect will make the deformation of the matrix more difficult.

The OCP values monitored over 24 h for the AZ91D coating and the SiC/AZ91D composite coating, as well as the magnesium substrate, are shown in Figure 9a. Obviously, the potential of magnesium immersed in this medium is low. This is why magnesium is always used as a sacrificial anode to protect metal installations. Both coatings show higher potential than the magnesium substrate. It is, thus, demonstrated that the intermetallic compound $Mg_{17}Al_{12}$ does not intervene in corrosion mechanisms after 24 h of immersion. The composite coating has a significant nobler potential. The densification effect related to the presence of hard SiC particles limits the microcavities involving in the crevice corrosion initiation [13], which facilitates the formation of the MgO oxide film.

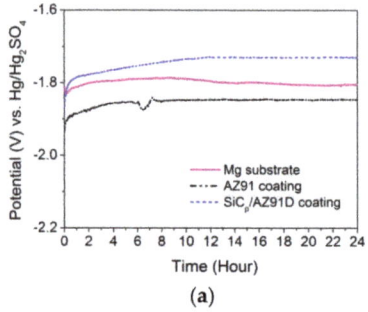 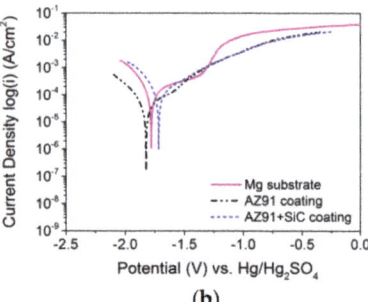

Figure 9. Corrosion characters for cold-sprayed AZ91D coating and SiC/AZ91D composite coating: open circuit potentials (**a**), and potentiodynamic polarization curves (**b**).

Figure 9b shows the polarization curves of the coatings compared to that of the substrate. These polarization curves complete the corrosion potential measurements that give the thermodynamic criteria only. Each range of the polarization curves gives information on the kinetics and the nature of the reactions that take place at the interfaces. The ranking of the corrosion potentials is the same as for Figure 9a. The corrosion potential of the Mg substrate is the same as the potential of the AZ91D coating. It is close to -1.9 V/Hg/Hg$_2$SO$_4$. For the composite coating this value is nobler and close to -1.75 V/Hg/Hg$_2$SO$_4$.

For potential values lower than the corrosion potentials, the polarization curves have the same trends and the same slopes. The recorded phenomena are identical. This is due to the reduction of hydrogen. Since the involved reactions are carried out by charge transfer, no diffusion of dissolved oxygen contribution is recorded. The significant difference in all electrochemical behaviors occurs for the anodic branch of the electrochemical behavior of the materials, for the nobler potentials, located after the corrosion potential. The substrate has a slight passive domain between -1.9 to -1.5 V/Hg/Hg$_2$SO$_4$, for which the current is relatively low. Beyond -1.5 V/Hg/Hg$_2$SO$_4$, the current density increases significantly to reach a diffusion plateau relative to the generalized corrosion of Mg. This polarization curve indicates two points: (i) interest of the polarization curves that permit to indicate the corrosion mode of metallic materials and (ii) Mg is very reactive in this medium. Indeed, if the AZ91D corrosion potential is identical to that of Mg, it should be noted that its electrochemical behavior in the anodic domain is very different. The shape of the curve is essentially the same as that of Mg, which is normal since the main element of AZ91D is magnesium. However, the current densities are much lower. This result is due to the presence of aluminum in the alloy and also to the coating morphology which has no porosity. This porosity could, indeed, show localized corrosion phenomena by the effect of cracks, which would be characterized by significant current densities increasing. The kinetics recorded for this coating is closed to that recorded for a passivated coating, thus, it is less reactive than the substrate. The electrochemical behavior of the composite coating is identical to that of the coating prepared with AZ91D alone. The addition of SiC powder particles does not affect the electrochemical behavior of the AZ91D coating, it only contributes to the densification of the coatings, which had already been demonstrated in a previous study dedicated to another sacrificial system [12].

Measurements using a zero resistance ammeter device were undertaken to analyze the galvanic couplings between the substrate and the coatings. These measurements give results to the mixed potential and the mixed current of the galvanic coupling. These results are presented in Figure 10. The measurements were carried out for 50 h. The mixed potential measured between AZ91D and the substrate is identical to that of the two materials, measured separately. The potential difference between these two elements is close to zero, which explains the value of the potential obtained, and which also explains why the galvanic current recorded for this AZ91D-Mg substrate system is close to zero, too. The AZ91D coating and the magnesium substrate show similar corrosion potentials in polarization tests, which results in a difficulty of predicting the polarity of this galvanic couple. The direction of galvanic current ensures that AZ91D coating is nobler than the magnesium substrate. The couple of SiC/AZ91D coating-Mg substrate shows higher galvanic potential and more negative galvanic current than the couple of AZ91D coating-Mg substrate. The mixed potential of the SiC/AZ91D-Mg substrate is driven by the potential of coating which was presented in Figure 10. The addition of SiC powder particles increase the absolute value of galvanic current, i.e., the SiC/AZ91D coating is nobler than the AZ91D coating on the Mg substrate.

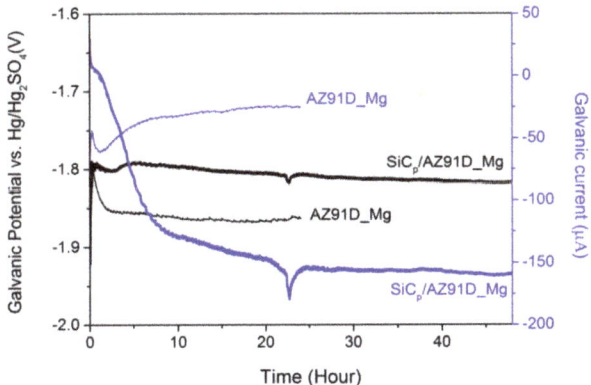

Figure 10. Time behavior of galvanic potential and galvanic current for the AZ91D coating and the SiC/AZ91D composite coating coupled to the magnesium substrate.

Electrochemical measurements confirm that, in both cases recorded in Figures 9 and 10, the behavior of the coatings controls the electrochemical behavior of the AZ91D-Mg system with or without SiC. In the case of an AZ91D coating, the galvanic coupling will not be effective because the potential difference and the galvanic current are close to zero. In the case of a SiC/AZ91D-Mg coating, the measurements indicate that when the composite coating is used to protect the Mg substrate, the galvanic coupling does not exist since the porosity is low. The addition of SiC during the spraying makes it possible to densify the coating and, thus, to avoid the connective porosity through the substrate.

4. Conclusions

AZ91D coatings and SiC-reinforced AZ91D composite coating were fabricated using cold spraying. Compressed air was used as the main gas, and the effects of gas temperature were studied. It was found that the gas temperature should be higher than 450 °C for cold spraying of spherical AZ91D powder particles. The deposition efficiency increases greatly with the gas temperature and it is lower than 10% for all coating specimens. The coating average porosity remained between 3.6% and 3.9% when the gas temperature is in the range of 500~600 °C. Coatings are denser when the gas temperature is higher. XRD results showed that no phase transformation or oxidization occurred during the cold spraying process of AZ91D. The microhardness of the coatings (approximately 100 HV) was much higher than that of as-casted bulks due to the work hardening effect. Gas temperature does not show an obvious influence on microhardness. Partial impact melting was found between AZ91D particles in the coatings fabricated at 550 °C and 600 °C.

15 vol % SiC powder particles were added to feedstock powder as reinforcement to improve the coating performance. Results showed that the SiC content of coatings is 10.7 ± 3.2 vol %. Reduced SiC content means that the deposition efficiency of the SiC powder particles is lower than the deposition efficiency of the AZ91D particles. The addition of SiC particles reduces the porosity and increases the microhardness of cold-sprayed AZ91D coatings. Both the AZ91D coating and SiC-reinforced AZ91D composite coating serve as the cathode for the magnesium substrate due to the relatively higher corrosion potential. The addition of SiC particles increases the open circuit potential of the AZ91D coating, meanwhile increasing the galvanic potential and decreasing the negative galvanic current of the coating-substrate couple.

Acknowledgments: This investigation is financially supported by the National Natural Science Foundation of China under grant no. 51601158 and the Qihang Science Research Foundation of Southwest Petroleum University, no. 2015QHZ013.

Author Contributions: X.S., M.P., and H.L. conceived and designed the experiments; Y.W. and X.S. performed the experiments; X.S., Y.W., and J.T. analyzed the data; B.N., M.P., and H.L. contributed materials; J.T. contributed analysis tools; and Y.W., X.S., and B.N. wrote the paper.

Conflicts of Interest: The authors declare no conflict of interest.

References

1. Esmaily, M.; Svensson, J.E.; Fajardo, S.; Birbilis, N.; Frankel, G.S.; Virtanen, S.; Arrabal, R.; Thomas, S.; Johansson, L.G. Fundamentals and Advances in Magnesium Alloy Corrosion. *Prog. Mater. Sci.* **2017**, *89*, 92–193. [CrossRef]
2. Liu, F.; Zhang, J.; Sun, C.; Yu, Z.; Hou, B. The Corrosion of Two Aluminium Sacrificial Anode Alloys in SRB-containing Sea Mud. *Corros. Sci.* **2014**, *83*, 375–381. [CrossRef]
3. Lun Sin, S.; Dubé, D.; Tremblay, R. An Investigation on Microstructural and Mechanical Properties of Solid Mould Investment Casting of AZ91D Magnesium Alloy. *Mater. Charact.* **2008**, *59*, 178–187. [CrossRef]
4. Kocurek, R.; Adamiec, J. The Repair Welding Technology of Casts Magnesium Alloy QE22. *Solid State Phenom.* **2013**, *212*, 81–86. [CrossRef]
5. Sun, D.X.; Cui, D.L.; Shi, J.T. Hot Cracking and Microstructure of Welding Joint of Magnesium Alloy AZ91D. *Adv. Mater. Res.* **2013**, *753–755*, 435–438. [CrossRef]
6. Ye, H.; Zhang, X.B.; Chang, X.; Chen, R. Microstructures and Properties of Laser Al Alloying on AZ31 Magnesium Alloy. *Adv. Mater. Res.* **2011**, *189–193*, 867–870. [CrossRef]
7. Suo, X.; Guo, X.; Li, W.; Planche, M.-P.; Liao, H. Investigation of Deposition Behavior of Cold-Sprayed Magnesium Coating. *J. Therm. Spray Technol.* **2012**, *21*, 831–837. [CrossRef]
8. Suo, X.; Guo, X.; Li, W.; Planche, M.-P.; Bolot, R.; Liao, H.; Coddet, C. Preparation and Characterization of Magnesium Coating Deposited by Cold Spraying. *J. Mater. Process. Technol.* **2012**, *212*, 100–105. [CrossRef]
9. Li, C.; Li, W.; Liao, H. Examination of the Critical Velocity for Deposition of Particles in Cold Spraying. *J. Therm. Spray Technol.* **2006**, *15*, 212–222. [CrossRef]
10. Pathak, S.; Saha, G.C. Development of Sustainable Cold Spray Coatings and 3D Additive Manufacturing Components for Repair/Manufacturing Applications: A Critical Review. *Coatings* **2017**, *7*, 122. [CrossRef]
11. Ren, Y.Q.; King, P.C.; Yang, Y.S.; Xiao, T.Q.; Chu, C.; Gulizia, S.; Murphy, A.B. Characterization of heat treatment-induced pore structure changes in cold-sprayed titanium. *Mater. Charact.* **2017**, *132*, 69–75. [CrossRef]
12. Wang, Y.; Normand, B.; Mary, N.; Yu, M.; Liao, H. Microstructure and corrosion behavior of cold sprayed SiCp/Al 5056 composite coatings. *Surf. Coat. Technol.* **2014**, *251*, 264–275. [CrossRef]
13. Wang, Y.; Normand, B.; Mary, N.; Yu, M.; Liao, H. Effects of ceramic particle size on microstructure and the corrosion behavior of cold sprayed SiCp/Al 5056 composite coatings. *Surf. Coat. Technol.* **2017**, *315*, 314–325. [CrossRef]
14. Villafuerte, J. Practical Cold Spray Success Repair of Al and Mg Alloy Aircraft Components. *Int. Therm. SPRAY Surf. Eng.* **2010**, *5*, 53–55.
15. Spencer, K.; Fabijanic, D.M.; Zhang, M.-X. The use of Al–Al$_2$O$_3$ cold spray coatings to improve the surface properties of magnesium alloys. *Surf. Coat. Technol.* **2009**, *204*, 336–344. [CrossRef]

16. Yoon, S.; Kim, H.; Lee, C. Fabrication of automotive heat exchanger using kinetic spraying process. *Surf. Coat. Technol.* **2007**, *201*, 9524–9532. [CrossRef]
17. Rech, S.; Surpi, A.; Vezzù, S.; Patelli, A.; Trentin, A.; Glor, J.; Frodelius, J.; Hultman, L.; Eklund, P. Cold-spray deposition of Ti 2 AlC coatings. *Vacuum* **2013**, *94*, 69–73. [CrossRef]
18. Rahim, T.A.; Takahashi, K.; Yamada, M.; Fukumoto, M. Effect of Powder Calcination on the Cold Spray Titanium Dioxide Coating. *Mater. Trans.* **2016**, *57*, 1345–1350. [CrossRef]
19. Chang, L.L.; Shang, E.F.; Wang, Y.N.; Zhao, X.; Qi, M. Texture and microstructure evolution in cold rolled AZ31 magnesium alloy. *Mater. Charact.* **2009**, *60*, 487–491. [CrossRef]
20. Yu, M. Elaboration de Composites à Matrice Métallique D'alliages D'aluminium par Projection à Froid. Ph.D. Thesis, Université de Technologie de Belfort-Montbéliard Ecole, Belfort, France, 2013.
21. Sumida, M.; Jung, S.; Okane, T. Solidification Microstructure, Thermal Properties and Hardness of Magnesium Alloy 20 mass% Gd Added AZ91D. *Mater. Trans.* **2009**, *50*, 1161–1168. [CrossRef]
22. Van Steenkiste, T.H.; Smith, J.R.; Teets, R.E.; Moleski, J.J.; Gorkiewicz, D.W.; Tison, R.P.; Marantz, D.R.; Kowalsky, K.A.; Riggs, W.L.; Zajchowski, P.H.; et al. Kinetic spray coatings. *Surf. Coat. Technol.* **1999**, *111*, 62–71. [CrossRef]
23. Suo, X.K.; Guo, X.P.; Li, W.Y.; Planche, M.P.; Zhang, C.; Liao, H.L. Microstructure and wear behavior of SiCp-reinforced magnesium matrix composite by cold spraying. In *Advanced Manufacturing Technology, Pts 1–3, Guangzhou, China, 16–18 September, 2011*; Gao, J., Ed.; TRANS TECH PUBLICATIONS Ltd.: Zurich, Switzerland, 2011; Volume 314–316, pp. 253–258.

© 2018 by the authors. Licensee MDPI, Basel, Switzerland. This article is an open access article distributed under the terms and conditions of the Creative Commons Attribution (CC BY) license (http://creativecommons.org/licenses/by/4.0/).

Article

Preparation of Metal Coatings on Steel Balls Using Mechanical Coating Technique and Its Process Analysis

Liang Hao [1,2], Hiroyuki Yoshida [3], Takaomi Itoi [4] and Yun Lu [4,*]

1. Tianjin Key Lab. of Integrated Design and On-Line Monitoring for Light Industry & Food Machinery and Equipment, Tianjin 300222, China; haoliang@tust.edu.cn
2. College of Mechanical Engineering, Tianjin University of Science and Technology, No. 1038, Dagu Nanlu, Hexi-District, Tianjin 300222, China
3. Chiba Industrial Technology Research Institute, 6-13-1, Tendai, Inage-ku, Chiba 263-0016, Japan; h.yshd14@pref.chiba.lg.jp
4. College of Mechanical Engineering & Graduate School, Chiba University, 1-33, Yayoi-cho, Inage-ku, Chiba 263-8522, Japan; itoi@chiba-u.jp
* Correspondence: luyun@faculty.chiba-u.jp; Tel.: +81-43-290-3514

Academic Editors: Tony Hughes and Russel Varley
Received: 1 February 2017; Accepted: 7 April 2017; Published: 10 April 2017

Abstract: We successfully applied mechanical coating technique to prepare Ti coatings on the substrates of steel balls and stainless steel balls. The prepared samples were analyzed by X-ray diffraction (XRD) and scanning electron microscopy (SEM). The weight increase of the ball substrates and the average thickness of Ti coatings were also monitored. The results show that continuous Ti coatings were prepared at different revolution speeds after different durations. Higher revolution speed can accelerate the formation of continuous Ti coatings. Substrate hardness also markedly affected the formation of Ti coatings. Specifically, the substance with lower surface hardness was more suitable as the substrate on which to prepare Ti coatings. The substrate material plays a key role in the formation of Ti coatings. Specifically, Ti coatings formed more easily on metal/alloy balls than ceramic balls. The above conclusion can also be applied to other metal or alloy coatings on metal/alloy and ceramic substrates.

Keywords: Ti coatings; steel balls; mechanical coating; process analysis

1. Introduction

Coating technology is one of the most frequently-used surface modification technologies, and has been applied in many engineering fields, including corrosion prevention [1], thermal barrier [2,3], anti-friction [4,5], stealth materials [6], etc. Other functions such as photocatalytic activity have also been found in metal/alloy coatings after certain treatments including thermal oxidation [7,8], chemical oxidation [9,10], plasma electrolytic oxidation [11,12], anodic oxidation [13,14], among others. In our published work, we prepared TiO_2/Ti composite photocatalyst coatings on the substrate of Al_2O_3 balls using mechanical coating followed by thermal oxidation [15]. With further study, we developed oxygen-deficient visible-light-responsive TiO_2 coatings [16]. Therefore, the preparation of metal/alloy coatings is of paramount practical importance. Researchers have prepared several kinds of metal/alloy coatings on ceramic or metal substrates using mechanical coating technique [17]. Early in 1995, Kobayashi developed Al and Ti-Al coatings on the substrates of stainless steel balls and ZrO_2 balls [18]. Romankov et al. [19] also prepared Al and Ti-Al coatings on a Ti alloy substrate. Gupta et al. [20] prepared nanocrystalline Fe-Si alloy coatings on a mild steel substrate. Farahbakhsh et al. [21] deposited Cu and Ni-Cu solid solution coatings on ceramic and metal substrates. We have fabricated Fe and Zn coatings on Al_2O_3 ball substrates [22,23]. Furthermore, we have also revealed that the properties of the metal powder played an important role in the formation of metal coatings [24]. Besides the influence of some processing parameters including milling speed and time, a possible mechanism of coatings' formation was further studied in [20,21,23].

However, the influence of substrates including material properties and surface roughness on the formation of metal coatings has not been revealed so far. In this work, we would verify the formation possibility of metallic coatings on metallic substrates and attempted to prepare Ti coatings on different steel substrates utilizing mechanical coating technique. The formation process of Ti coatings and the influence of substrates' properties on their formation were also involved.

2. Materials and Methods

Ti coatings were prepared using a mechanical coating technique with a planetary ball mill (Pulverisette 6, Fritsch). The transmission ratio of the mill was 1:−1.82. Ti powder (Osaka Titanium Technologies Co. Ltd., Osaka, Japan) and steel balls as the substrates were charged into a bowl made of alumina (volume: 250 mL). The bowl was fixed in the planetary ball mill, and then the mechanical coating process was carried out at different rotational speeds for different durations. Two kinds of substrates were used separately to clarify the influence of steel ball substrates on the formation of metal coatings, including steel balls (SUJ-2, density of 7.85 $g \cdot cm^{-3}$) and stainless steel balls (SUS-304, density of 7.93 $g \cdot cm^{-3}$). The composition of steel (SUJ-2) and stainless steel balls (SUS-304) is listed in Table 1. To study the influence of substrates' surface roughness, steel balls were polished to make their surface smoother before the mechanical coating process. The polishing process is as follows. Firstly, abrasive paper with mesh number of 80 was put into the bowl along the wall of the bowl. Secondly, the balls were charged into the bowl and ball milling was carried out. In the ball milling, the balls were polished by the abrasive paper throughout their repeated collision and friction with the ball of the bowl. The surface roughness of the balls was not measured. Meanwhile, steel balls were annealed in vacuum at 1073 K holding for 1.5 h to change their hardness before mechanical coating process to study the influence of the substrate hardness on the coatings' formation. Tables 2 and 3 give the relevant processing parameters and the sample symbols. The average particle size distribution of titanium powder is about 30 μm, ranging from 5–100 μm. Most of them (up to 70%) are located in the range of 20–50 μm. The parameters x and y correspond to the rotational speed of the mill and milling time, respectively. The volume ratio of metallic powder to the balls and the filling degree are 1:1.7 and 5%, respectively.

Table 1. Composition of steel (SUJ-2) and stainless steel (SUS-304) balls in the work.

No.	C	Si	Mn	P	S	Ni	Cr	Others
SUJ-2	0.95–1.10	0.15–0.35	≤0.50	≤0.025	≤0.025	–	1.30–1.60	Fe
SUS-304	≤0.08	≤1.00	≤2.00	≤0.045	≤0.030	8.00–10.50	18.00–20.00	Fe

Table 2. Relevant processing parameters in the present work.

	Raw Materials	Weight (g)	Average Diameter (mm)	Purity (%)
Metal powder	Ti powder	20.0	0.03	99.1
Substrates	Steel balls	58.5	1.0	SUJ-2
	Stainless steel balls	59.5	1.0	SUS-304

Table 3. Relevant sample symbols and treatment condition of the balls in the present work.

Sample Symbol	Substrate	Surface Roughness	Hardness (HV)
TSx-y	SUJ-2	original	809
TSSx-y	SUJ-2	polished	809
TSYx-y	SUJ-2	original	201
TBx-y	SUS-304	original	187

All samples were characterized by X-ray diffractometer (XRD) (JDX-3530, JEOL, Tokyo, Japan) with Cu Kα radiation at 30 kV and 20 mA to determine the phases present. A scanning electron microscope (SEM) (JSM-6510A, JEOL, Tokyo, Japan) was used to observe the surface morphologies and the microstructure of the cross-sections of the Ti-coated steel balls. The average thickness of Ti coatings was estimated from 40 different locations of five Ti-coated steel balls in their SEM images of the cross sections. The average weight increase of 50 steel balls during mechanical coating process was also calculated by weighing 50 randomly-selected Ti-coated steel balls three times.

3. Results and Discussion

3.1. Preparation of Ti Coatings on Steel Balls

The XRD patterns of the Ti-coated steel balls are presented in Figure 1. We could see the diffraction peaks of Ti in addition to those of Fe from the XRD patterns when the duration of mechanical coating processing was increased to 4 and 8 h. This means that some Ti powder particles adhered to the surface of the steel balls. When processing time reached 10 h, the diffraction peaks of Fe could no longer be observed, indicating that continuous Ti coatings had formed on the steel balls.

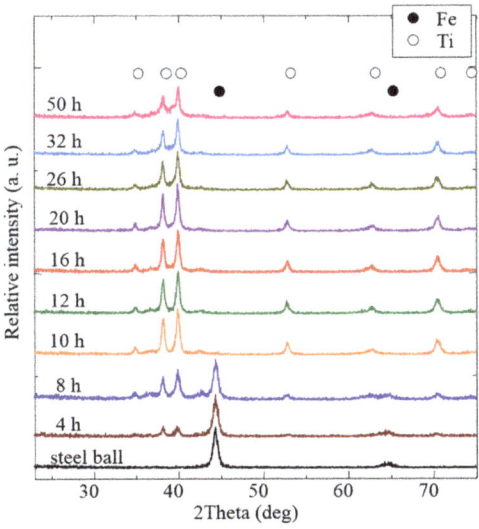

Figure 1. XRD patterns of the Ti coatings prepared by mechanical coating on steel balls at 300 rpm.

The surface morphologies of Ti-coated steel balls were recorded by SEM and are displayed in Figure 2. When the duration of the mechanical coating process was 4 and 8 h, Ti powder particles discontinuously coated the surfaces of steel balls (Figure 2a,b). With the increase of process duration to 10 h, continuous Ti coatings formed (Figure 2c). The surface of the Ti coatings became rugged, and humps were formed with further increase of duration to 50 h. The results from the SEM images are consistent with that reflected from the XRD patterns in Figure 1. Figure 3 shows the SEM images of the cross-section of the samples prepared by mechanical coating. Although the coating of Ti powder particles on steel balls was not clearly observed (Figure 3a,b) when the duration was 4 or 8 h, the formation of continuous Ti coatings was confirmed after 8 h of mechanical coating (Figure 3c). The coatings' evolution in Figure 3 agreed with that in Figures 1 and 2. Therefore, we can say that continuous Ti coatings on steel balls were prepared at 300 rpm after 10 h of mechanical coating process from the above results.

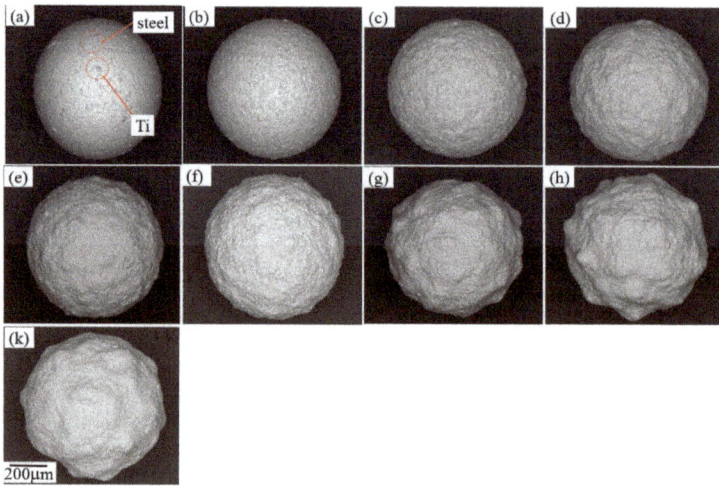

Figure 2. SEM images for the surface morphologies of the samples prepared by mechanical coating at 300 rpm after different duration: (**a**) 4 h; (**b**) 8 h; (**c**) 10 h; (**d**) 12 h; (**e**) 16 h; (**f**) 20 h; (**g**) 26 h; (**h**) 32 h; and (**k**) 50 h.

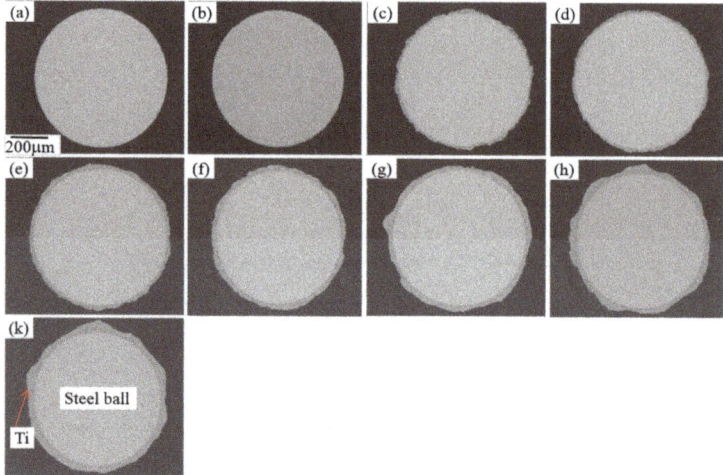

Figure 3. SEM images for the cross-section of the samples prepared by mechanical coating at 300 rpm after different durations: (**a**) 4 h; (**b**) 8 h; (**c**) 10 h; (**d**) 12 h; (**e**) 16 h; (**f**) 20 h; (**g**) 26 h; (**h**) 32 h; and (**k**) 50 h.

3.2. Influence of Rotational Speed

To study the influence of revolution speed on the formation of Ti coatings, continuous Ti coatings on steel balls were also prepared at 400 rpm, with the results displayed in Figure 4. We can see that continuous Ti coatings have been formed after 4 h of the mechanical coating process (Figure 4b). The thickness of the Ti coatings was increased with the increase of duration from 4 h to 20 h. However, continuous Ti coatings began to separate from steel balls as the duration was further increased to 26 and 32 h. Therefore, we can say the evolution includes the following four stages: nucleation, growth of nuclei, formation of coatings, and exfoliation. The evolution is similar to that of Fe and Zn coatings [22,23].

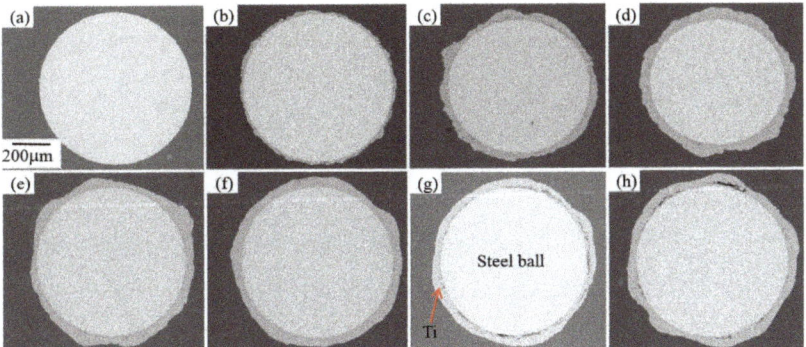

Figure 4. SEM images for cross-section of Ti samples prepared by mechanical coating at 400 rpm after different durations: (**a**) 1 h; (**b**) 4 h; (**c**) 8 h; (**d**) 12 h; (**e**) 16 h; (**f**) 20 h; (**g**) 26 h; and (**h**) 32 h.

The weight increase of 50 steel balls during mechanical coating at different revolution speeds was recorded as illustrated in Figure 5. The weight increase means that more Ti powder particles coat the steel balls. We found that the weight of the steel balls increased with the increase in duration. However, the weight increase became greater with the increase of revolution speed from 200 to 400 rpm at the same mechanical coating process duration. This suggests that a higher revolution speed can accelerate the coating of Ti powder particles on the surface of steel balls. The average thickness change of continuous Ti coatings was also monitored as shown in Figure 6. We can note that the data at 200 rpm is absent because continuous Ti coatings were not even successfully prepared after 50 h. This hints that continuous Ti coatings may not be formed at revolution speeds of 200 rpm or lower. The average thickness evolution of continuous Ti coatings at 300 and 400 rpm is similar to the weight increase change in Figure 5. When rotational speed was 400 rpm, the weight began to decrease when the time came to 26 h, as the formed coatings began to peel off. If milling time is prolonged any further, the exfoliation of metallic coatings will continue. Therefore, we did not provide data after 26 h. According to the parameters named "collision strength" and "collision power" which we proposed in published work [23], the energy transferred to the metallic powder particles from the balls quickly increases with the increase of rotation speed of the ball mill. Greater collision power means larger transferred collision energy, which creates severe plastic deformation. The cold welding among metallic powder particles occurs only when plastic deformation is greater than a critical value [24].

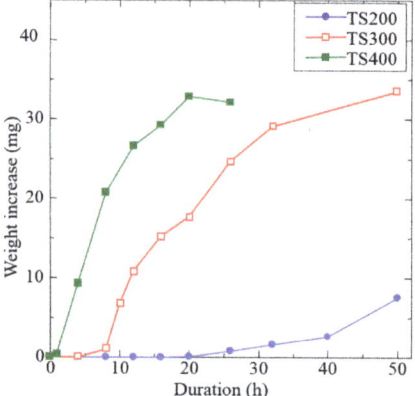

Figure 5. Weight increase of 50 steel balls versus duration of mechanical coating at different revolution speeds.

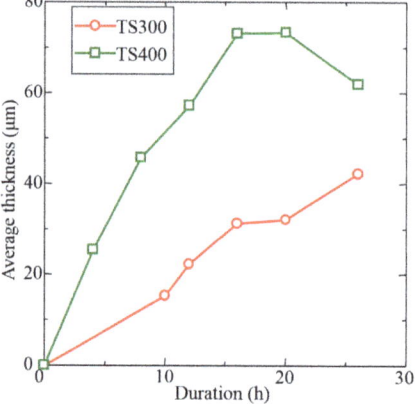

Figure 6. Average thickness of Ti coatings versus duration of mechanical coating at different revolution speeds.

3.3. Influence of Surface Roughness and Hardness

The SEM images for the morphologies of the samples prepared by the mechanical coating process are shown in Figure 7. Figure 7a,b show the influence of surface roughness on the coatings' formation. No evident difference can be observed from the SEM images. From Figure 8, we can see that the initial coating rate for the polished balls was slightly greater than those which were unpolished. In other words, the decrease in surface roughness favors the adhesion of metallic powder particles to the surface of metallic balls. We believe that the surface roughness improvement can decrease the air volume reserved in the cavities in the surface of the balls. The contact area among the balls and the metallic powder particles was increased, which can increase the possibility of cold welding. Therefore, the surface roughness improvement accelerated the formation of metallic coatings. On the other hand, surface roughness improvement decreased the quantity of the cavities in the surface. Therefore, the interaction opportunity—specifically the mechanical inter-locking between the cavities and the metallic particles—was decreased. Finally, the surface roughness improvement hinders the formation of metallic coatings. According to the above results, we can conclude that the influence of the surface roughness on the formation of metallic coatings is rather complex; the coexistence of promoting and obstructive factors made the influence negligible. As for the influence of the substrates' hardness, the formation situation of Ti coatings is given in Figure 7c,d. We can clearly see that more Ti powder particles were adhered to the annealed steel ball than to the steel ball. In other words, Ti powder particles more easily coat the softer steel balls. A slight difference in weight increase shown in Figure 8 also proved this. The influence of balls' surface hardness on the coating of metallic powder particles can also be attributed to the cold welding of metallic powder. As discussed above, the cold welding among balls and metallic powder particles happens only when a critical plastic strain is satisfied. After they were annealed, the balls became softer than that before annealing. During the collision among balls and metallic powder particles, the softer surface of the balls welds with the metallic particles more easily. After the surface of these balls is totally coated with metallic powders after 12 h of ball milling, the interaction among balls and metallic powder particles has been replaced by that among metallic powder particles. Therefore, the influence of surface parameters including roughness and hardness cannot be studied any more.

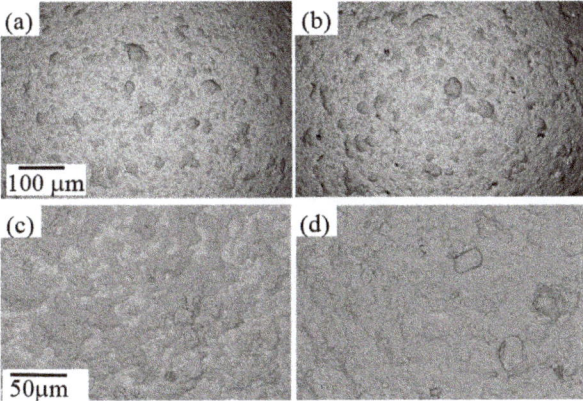

Figure 7. SEM images of morphologies of the samples: (**a**) TS300-4; (**b**) TSS300-4; (**c**) TS300-8; and (**d**) TSY300-8.

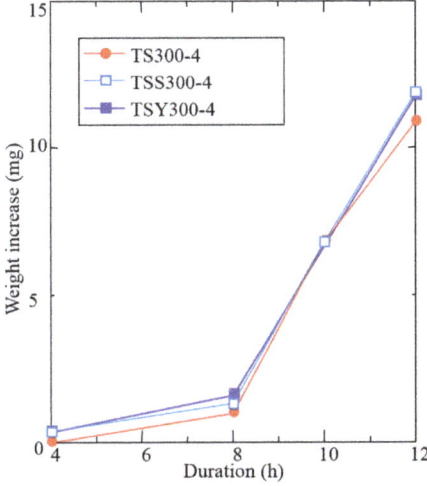

Figure 8. Weight increase of 50 steel balls during mechanical coating versus duration.

3.4. Influence of Substrate Material

We also studied the influence of substrate material on the formation of continuous Ti coatings. When stainless steel (SUS-304) balls were chosen as the substrate (Figure 9), the formation of continuous Ti coatings took about 10 h, which was identical to that using steel (SUJ-2) balls as the substrate. This means that the required time to form continuous Ti coatings on stainless steel and steel balls was identical. However, the formation of continuous Ti coatings on Al_2O_3 balls in the same condition took 20 h [25]. Therefore, we can conclude that the formation of Ti coatings on steel balls is easier and quicker than on ceramic balls. In other words, the substrate plays a key role in the formation of Ti coatings. From the above results, we can state that the formation of Ti coatings on steel balls is much easier and quicker than on ceramic balls. The influence of substrate material on the formation of metallic coatings can be explained as follows.

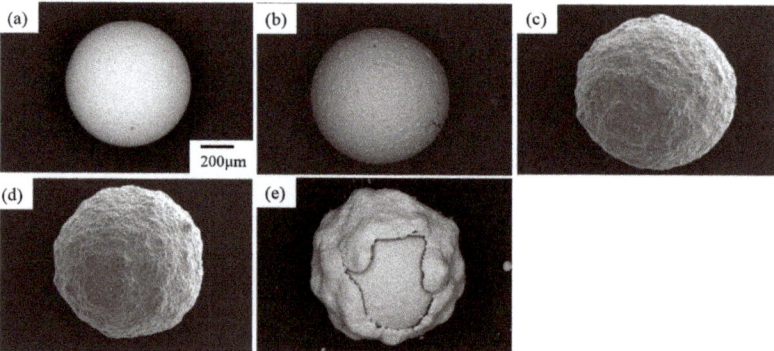

Figure 9. SEM images of surface morphologies of Ti coatings on stainless steel (SUS-304) balls prepared by mechanical coating at 300 rpm after different duration: (**a**) 0 h; (**b**) 4 h; (**c**) 10 h; (**d**) 16 h and (**e**) 60 h.

Firstly, it is well-known that the hardness of Al_2O_3 is about 2300 HV, which is far greater than those of steel and stainless steel. When substrate hardness decreased from 2300 HV to 809 HV, the required time to form continuous Ti coatings decreased from 20 h to 10 h. With a further decrease of substrate hardness from 809 HV to 187 HV, the required time hardly decreased, indicating that decreasing the substrates' hardness within a certain range can shorten the formation of Ti coatings. The influence of substrate hardness has been discussed above. Secondly, the material transfer from metal to ceramics is difficult than cold welding between metal materials [26]. Some works [24,26] have proved that the interaction between metallic particles and ceramic balls belongs to mechanical self-locking due to the plastic deformation of metallic particles. However, the cold welding occurred among the fresh surface of the balls and metallic powder particles when balls became metallic ones. The strength of self-locking is lower than that of cold welding.

4. Conclusions

Continuous Ti coatings on steel and stainless steel ball substrates were prepared by mechanical coating technique. Greater revolution speed, providing larger collision force and energy, accelerated the formation of continuous Ti coatings. The substrate material plays an essential role in the formation of Ti coatings; specifically, Ti coatings or even other metal coatings are more easily formed on metal/alloy balls than on ceramic balls. Meanwhile, substrate hardness also markedly affected the formation of Ti coatings. The material with smaller surface hardness is more suitable as the substrate on which Ti coatings were prepared. The above conclusion can also be exerted on other metal or alloy coatings on metal/alloy and ceramic substrates.

Acknowledgments: This work is financially supported by the National Nature Science Foundation of China (No. 51404170), the Innovation Team Program of Tianjin University of Science & Technology (No. 10117) and the Scientific Research Foundation of Tianjin University of Science & Technology (No. 10220).

Author Contributions: Y. Lu conceived and designed the experiments; L. Hao performed the experiments; L. Hao and H. Yoshida analyzed the data; T. Itoi contributed analysis tools; L. Hao wrote the paper.

Conflicts of Interest: The authors declare no conflict of interest.

References

1. Ramanauskas, R.; Quintana, P.; Maldonado, L.; Pomés, R.; Pech-Canul, M.A. Corrosion resistance and microstructure of electrodeposited Zn and Zn alloy coatings. *Surf. Coat. Technol.* **1997**, *92*, 16–21. [CrossRef]
2. Tang, J.J.; Bai, Y.; Zhang, J.C.; Liu, K.; Liu, X.Y.; Zhang, P.; Wang, Y.; Zhang, L.; Liang, G.Y.; Gao, Y.; et al. Microstructural design and oxidation resistance of CoNiCrAlY alloy coatings in thermal barrier coating system. *J. Alloys. Compd.* **2016**, *688*, 729–741. [CrossRef]
3. Zhou, C.; Zhang, Q.; Li, Y. Thermal shock behavior of nanostructured and microstructured thermal barrier coatings on a Fe-based alloy. *Surf. Coat. Technol.* **2013**, *217*, 70–75. [CrossRef]
4. Yamamoto, K.; Ito, H.; Kujime, S. Nano-multilayered CrN/BCN coating for anti-wear and low friction applications. *Surf. Coat. Technol.* **2007**, *201*, 5244–5248. [CrossRef]
5. Wei, S.; Pei, X.; Shi, B.; Shao, T.; Li, T.; Li, Y.; Xie, Y. Wear resistance and anti-friction of expansion cone with hard coating. *Petrol. Explor. Dev.* **2016**, *43*, 326–331. [CrossRef]
6. Wang, K.; Wang, C.; Yin, Y.; Chen, K. Modification of Al pigment with graphene for infrared/visual stealth compatible fabric coating. *J. Alloys Compd.* **2017**, *690*, 741–748. [CrossRef]
7. Yoshida, H.; Lu, Y.; Nakayama, H.; Hirohashi, M. Fabrication of TiO_2 film by mechanical coating technique and its photocatalytic activity. *J. Alloys Compd.* **2009**, *475*, 383–386. [CrossRef]
8. Khosravani, S.; Dehaghi, S.B.; Askari, M.B.; Khodadadi, M. The effect of various oxidation temperatures on structure of Ag-TiO_2 thin film. *Microelectron. Eng.* **2016**, *163*, 67–77. [CrossRef]
9. Sun, T.; Wang, M. Low-temperature biomimetic formation of apatite/TiO_2 composite coatings on Ti and NiTi shape memory alloy and their characterization. *Appl. Surf. Sci.* **2008**, *255*, 396–400.
10. Cotolan, N.; Rak, M.; Bele, M.; Cör, A.; Muresan, L.; Milošev, I. Sol-gel synthesis, characterization and properties of TiO_2 and Ag-TiO_2 coatings on titanium substrate. *Surf. Coat. Technol.* **2016**, *307A*, 790–799. [CrossRef]
11. He, J.; Luo, Q.; Cai, Q.Z.; Li, X.W.; Zhang, D.Q. Microstructure and photocatalytic properties of WO_3/TiO_2 composite films by plasma electrolytic oxidation. *Mater. Chem. Phys.* **2011**, *129*, 242–248. [CrossRef]
12. Stojadinović, S.; Radić, N.; Vasilić, R.; Petković, M.; Stefanov, P.; Zeković, L.; Grbić, B. Photocatalytic properties of TiO_2/WO_3 coatings formed by plasma electrolytic oxidation of titanium in 12-tungstosilicic acid. *Appl. Catal B Environ.* **2012**, *126*, 334–341. [CrossRef]
13. Ghicov, A.; Macak, J.M.; Tsuchiya, H.; Kunze, J.; Haeublein, V.; Frey, L.; Schmuki, P. Ion implantation and annealing for an efficient N-doping of TiO_2 nanotubes. *Nano Lett.* **2006**, *6*, 1080–1082. [CrossRef]
14. Schlott, F.; Ohser-Wiedemann, R.; Jordan, T.; Kreisel, G. Effect of the electrolyte composition on the anatase fraction of photocatalytic active TiO_2 coatings prepared by plasma assisted anodic oxidation. *Thin Solid Films* **2012**, *520*, 2549–2553. [CrossRef]
15. Lu, Y.; Matuszaka, K.; Hao, L.; Hirakawa, Y.; Yoshida, H.; Pan, F. Photocatalytic activity of TiO_2/Ti composite coatings fabricated by mechanical coating technique and subsequent heat oxidation. *Mater. Sci. Semicond. Proc.* **2013**, *16*, 1949–1956. [CrossRef]
16. Guan, S.; Hao, L.; Lu, Y.; Yoshida, H.; Pan, F.; Asanuma, H. Fabrication of oxygen-deficient TiO_2 coatings with nano-fiber morphology for visible-light photocatalysis. *Mater. Sci. Semicond. Proc.* **2016**, *41*, 358–363. [CrossRef]
17. Lu, Y.; Hirohashi, M.; Zhang, S. Fabrication of oxide film by mechanical coating technique. In Proceedings of the International Conference on Surface, Coatings and Nanostructured Materials, Aveiro, Portugal, 7–9 September 2005. Paper No. FP117.
18. Kobayashi, K. Formation of coating film on milling balls for mechanical alloying. *Mater. Trans.* **1995**, *36*, 134–137. [CrossRef]

19. Romankov, S.; Sha, W.; Kaloshkin, S.D.; Kaevitser, K. Formation of Ti-Al coatings by mechanical alloying method. *Surf. Coat. Technol.* **2006**, *201*, 3235–3245. [CrossRef]
20. Gupta, G.; Mondal, K.; Balasubramaniam, R. In situ nanocrystalline Fe-Si coating by mechanical alloy. *J. Alloys Compd.* **2009**, *482*, 118–122. [CrossRef]
21. Farahbakhsh, I.; Zakeri, A.; Manikandan, P.; Hokamoto, K. Evaluation of nanostructured coating layers formed on Ni balls during mechanical alloying of Cu powder. *Appl. Surf. Sci.* **2011**, *257*, 2830–2837. [CrossRef]
22. Hao, L.; Lu, Y.; Asanuma, H.; Guo, J. The influence of the processing parameters on the formation of iron thin films on alumina balls by mechanical coating technique. *J. Mater. Process. Technol.* **2012**, *212*, 1169–1176. [CrossRef]
23. Hao, L.; Lu, Y.; Sato, H.; Asanuma, H. Fabrication of zinc coatings on alumina balls from zinc powder by mechanical coating technique and the process analysis. *Powder Technol.* **2012**, *228*, 377–384. [CrossRef]
24. Lü, L.; Lai, M.; Zhang, S. Modeling of the mechanical-alloy process. *J. Mater. Process. Technol.* **1995**, *52*, 539–546. [CrossRef]
25. Lu, Y.; Guan, S.; Hao, L.; Yoshida, H. Review on the photocatalyst coatings of TiO_2: Fabrication by mechanical coating technique and its application. *Coatings* **2015**, *5*, 545–556. [CrossRef]
26. Hao, L.; Lu, Y.; Sato, H.; Asanuma, H.; Guo, J. Influence of metal properties on the formation and evolution of metal coatings during mechanical coating. *Metall. Mater. Trans. A* **2013**, *44*, 2717–2724. [CrossRef]

© 2017 by the authors. Licensee MDPI, Basel, Switzerland. This article is an open access article distributed under the terms and conditions of the Creative Commons Attribution (CC BY) license (http://creativecommons.org/licenses/by/4.0/).

Article

Characterization of Electroless Ni–P Coating Prepared on a Wrought ZE10 Magnesium Alloy

Martin Buchtík [1,*], Petr Kosár [1], Jaromír Wasserbauer [1], Jakub Tkacz [1] and Pavel Doležal [1,2]

1. Materials Research Centre, Faculty of Chemistry, Brno University of Technology, Purkyňova 464/118, 612 00 Brno, Czech Republic; xckosarp@fch.vut.cz (P.K.); wasserbauer@fch.vut.cz (J.W.); tkacz@fch.vut.cz (J.T.); dolezal@fme.vutbr.cz (P.D.)
2. Faculty of Mechanical Engineering, Brno University of Technology, Technická 2896/2, 602 00 Brno, Czech Republic
* Correspondence: xcbuchtik@fch.vut.cz; Tel.: +420-736-445-019

Received: 19 February 2018; Accepted: 6 March 2018; Published: 7 March 2018

Abstract: Electroless low-phosphorus Ni–P coating was deposited on a wrought ZE10 magnesium alloy including an advanced pre-treatment of the material surface before deposition. Uniform Ni–P coating with an average thickness of 10 μm was formed by 95.6 wt % Ni and 4.4 wt % P. The content of Ni and P was homogeneous in the entire cross-section of the coating. Applying the Ni–P coating to the magnesium substrate, the surface microhardness increased from 60 ± 4 HV 0.025 to 690 ± 30 HV 0.025. Using the scratch test, it was determined that deposited Ni–P coating exhibits a high degree of adhesion to the magnesium substrate. Electrochemical corrosion properties of Ni–P coating were analyzed using the polarization tests in 0.1 M NaCl, while the deposited Ni–P coating showed an improvement of the corrosion resistance when compared to the ZE10 magnesium alloy. Using the scanning electron microscopy analysis, it was determined that the fine morphology of the deposited Ni–P coating did not contain visible microcavities. The absence of macrodefects due to the adequate pre-treatment before coating was reflected on the mechanism of the coated ZE10 degradation in a 0.1 M NaCl solution.

Keywords: electroless deposition; Ni–P coating; magnesium alloy; ZE10; adhesion; microhardness; EDS analysis; polarization test

1. Introduction

Magnesium alloys are ranked among the lightest constructional metallic materials [1,2]. They find their application in the automotive and aerospace industry due to their low density and high value of specific strength, toughness, and good machinability [3]. Low corrosion resistance, high chemical reactivity, low hardness, and low wear and abrasion resistance are considerable disadvantages of magnesium-based materials [4].

Mg–Zn–RE-based magnesium alloys such as ZE10 and ZE41 achieve higher values of strength and better mechanical and corrosion properties in comparison with pure magnesium [2,5]. These alloys contain, besides Mg, Zn, and rare earth (RE) elements, Pr, Nd, La, Ce, and a small amount of Zr [1]. Zinc improves the strength and corrosion resistance of magnesium alloys. Rare earth elements improve the casting and mechanical properties (strength and creep resistance) of the alloys at higher temperatures. Zr is mainly used for grain refinement [1,6]. Although these alloys achieve better mechanical and corrosion properties in comparison with pure magnesium, their surface properties such as hardness, corrosion, and wear resistance are still inadequate for certain industrial applications.

There are several ways to improve magnesium alloys surface properties and resistivity, including galvanic or electroless deposited coatings, thermally sprayed coatings, and applications of conversion coatings [7,8].

One way to protect the material from corrosion, and improve the material surface's mechanical properties, is to apply electroless Ni–P coatings in a nickel bath [9]. Electroless deposited Ni–P coatings increase corrosion resistance as well as the surface's mechanical properties such as hardness and wear resistance. Applied low-phosphorus Ni–P coatings, compared with high-phosphorus Ni–P coatings, have a high value of hardness, a high density, and a high crystallinity [10]. However, the deposited high-phosphorus Ni–P coatings have higher corrosion resistance when compared to the low-phosphorus Ni–P coatings. Based on the phosphorus content in Ni–P coatings, low-, medium-, and high-phosphorus Ni–P coatings can be distinguished. Low-phosphorus Ni–P coatings contain 2–5 wt % phosphorus, medium-phosphorus Ni–P coatings contain 6–9 wt % phosphorus, and high-phosphorus Ni–P coatings contain >10 wt % phosphorus [9,11].

In [12], the corrosion behavior of three types of electroless deposited Ni–P coatings was studied. Ni–P coatings deposited on the mild steel contained 3.34% P (low phosphorus), 6.70% P (medium-phosphorus), and 13.30% P (high phosphorus). Based on the obtained results of potentiodynamic polarization tests and Nyquist plots of deposited Ni–P coatings in a 3.5% NaCl solution, it was determined that corrosion potentials and charge transfer resistances (in this case, equal to the polarization resistance) of deposited Ni–P coatings increased as phosphorus content increased. Corrosion potential E_{corr} of the low-phosphorus Ni–P coating obtained using the Tafel extrapolation method was −536 mV, and charge transfer resistance R_{ct} was 6.90 kΩ·cm^2. Corrosion potentials for medium- and high-phosphorus Ni–P coatings were −434 mV and −411 mV, respectively. Charge transfer resistances of medium- and high-phosphorus coatings were 24.86 kΩ·cm^2 and 37.45 kΩ·cm^2, respectively.

The adhesion of the coating to the substrate has a great influence on corrosion behavior and the mechanical properties of Ni–P coatings deposited on the substrate [10]. The adhesion of the Ni–P coating to the deposited substrate is significantly affected by appropriately selected pre-treatment of the substrate surface, such as grinding, blasting, degreasing, pickling, and interlayer deposition. [7,9,13]. A very frequent type of pre-treatment is zincate-zinc immersion [14–16], but the main disadvantage is the high pH value [14–17]. Zinc immersion is used to remove residual oxides and hydroxides from the surface of the substrate [14].

This paper deals with the preparation and characterization of an electroless deposited Ni–P coating deposited on a wrought ZE10 magnesium alloy. This study is focused on the measurement and evaluation of mechanical, physical, and corrosion properties of the coated substrate, while a specific pre-treatment was used before substrate coating. The mechanism of the corrosion attack and the corrosion resistance of the coated and plain material in 0.1 M NaCl were studied using potentiodynamic tests and immersion tests, and the results were analyzed in terms of light microscopy and scanning electron microscopy.

2. Materials and Methods

Wrought ZE10 magnesium alloy samples with dimensions of 20 × 20 × 1.6 mm^3 were used as a substrate material for deposition of the electroless Ni–P coating. The microstructure of the ZE10 magnesium alloy, shown in Figure 1, was characterized using the light microscopy (LM) and the individual microstructural features were identified using a Zeiss EVO LS-10 scanning electron microscope (SEM, Carl Zeiss Ltd., Cambridge, UK) with an energy-dispersive X-ray spectroscopy (EDS) Oxford Instruments Xmax 80 mm^2 detector (Oxford Instruments plc, Abingdon, UK) and AZtec software analysis (version 2.4). Elemental composition of the coated material shown in Table 1 obtained using the GDOES (glow-discharge optical emission spectroscopy) corresponds to the ASTM standard [18].

To reveal the ZE10 magnesium alloy microstructure, prepared and polished metallographic samples were poured into an acetic picral etchant (consisting of 4.2 g of picric acid, 10 mL of acetic acid, 10 mL of distilled water, and 70 mL of ethanol) for 3 s and then into 2% Nital for 1 s [19]. The microstructure of the substrate, characterized on the surface parallel to the material processing, was formed by solid solution grains of α-Mg, Mg$_7$Zn$_3$(RE) (La, Ce)-based particles and undissolved Zr particles (Figure 1b) [1,5]. Via EDS analysis, the presence of Ce, La, Pr, and Nd elements was found in the Mg$_7$Zn$_3$(RE) (La, Ce)-based particles.

Table 1. The elemental composition of the wrought ZE10 magnesium alloy (wt %) (GDOES).

Zn	Zr	Mn	Fe	Mg	Others
1.41	0.14	0.08	0.005	balance	max. 0.30

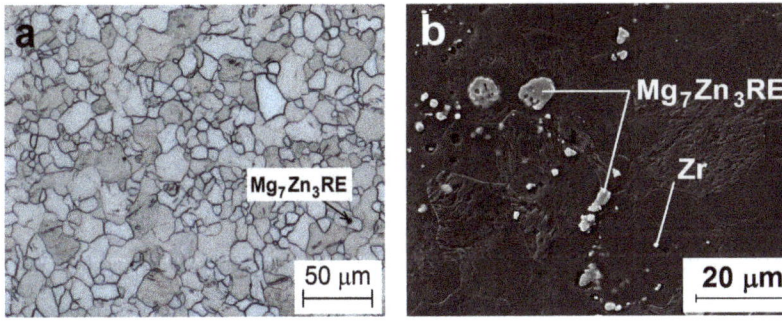

Figure 1. Surface microstructure of the ZE10 wrought magnesium alloy: (a) microstructure (LM); (b) intermetallic phase particles in microstructure (SEM).

To achieve a sufficient activity and roughness of the magnesium substrate surface before electroless Ni–P coating deposition, a suitably chosen specific pre-treatment is required. First, the samples were ground using SiC paper no. 1200 and then cleaned in an alkaline degreasing bath with the content of soil-releasing agents. To activate the magnesium substrate surface by partial etching, pickling in an acid pickling bath based in acetic acid was performed. Partial etching of the substrate surface led to an increase in the surface roughness and activity. This process was shown to have a positive influence on adhesion and mechanical properties of deposited Ni–P coating due to the mechanical interlocking of deposited Ni–P coating to magnesium substrate [20].

The electroless nickel bath contained a nickel source, NiSO$_4$·6H$_2$O, a reducing agent, NaH$_2$PO$_2$·H$_2$O, a complexing agent, and the substance-activating NaH$_2$PO$_2$ molecule. Samples were localized in the middle of the nickel bath to ensure uniform coating creation and the elimination of the thermal gradient from the bath surface [9]. The deposition of the electroless Ni–P coating proceeded for 60 min. For the microhardness measurement, a thicker coating was required, so a time of 180 min was taken for the coating preparation.

The distribution and the average content of nickel and phosphorus in deposited Ni–P coatings were determined using the Zeiss EVO LS-10 with an EDS Oxford Instruments Xmax 80 mm^2 detector SEM and the AZtec software (version 2.4). SEM observations were used to characterize Ni–P coating surface morphology and for the evaluation of the mechanism of corrosion degradation of the magnesium substrate with the subsequent violation of Ni–P coating after exposition in 0.1 M NaCl.

The microhardness measurement of the deposited Ni–P coating was carried out using the LECO AMH43 microhardness tester (Saint Joseph, MO, USA). The average value of the microhardness was obtained from 10 valid indentations performed within the coating cross section under the applied load of 25 g for 10 s, according to the ASTM E384 standard [21]. The indents were performed parallel to the coating surface.

Physical properties of the electroless deposited Ni–P coating were evaluated using the REVETEST scratch tester CSM Instruments with the Rockwell diamond indenter with a top angle of 120 and 200 μm radius hemispherical tip [22]. The progressive load type method was applied to the measurement.

The substrate surface was polished to the roughness of $R_a \approx 2$ μm using a DP-Paste M (diamond paste from Struers, Ballerup, Denmark) during the pre-treatment process before the alkaline degreasing in the addition. The friction force, the friction coefficient, the penetration depth, and the acoustic emission were recorded during the scratch test. Normal force was recorded. The applied normal force was set up in the range from 1 to 20 N. The speed of indenter was 1.58 mm·min^{-1}, and the total length of the trace was 3 mm.

The electrochemical corrosion characteristics of the ZE10 magnesium alloy and material with deposited Ni–P coating were analyzed in a 0.1 M solution of NaCl using the Bio-Logic VSP-300 potentiostat/galvanostat (BioLogic, Seyssinet-Pariset, France). Electrochemical polarization tests were performed on three specimens. A standard three-electrode cell was used for the measurements: a Pt gauze was used as a counter-electrode, a saturated calomel electrode (SCE) as a reference electrode, and a prepared sample as a working electrode. The analyzed sample area exposed to the solution during the polarization test was 1 cm^2. The stabilization time of the samples exposed to the corrosive environment was 5 min. The polarization range of the measurements was from −50 to +200 mV vs. open circuit potential (OCP). The corrosion potential E_{corr} and the corrosion current density i_{corr}, were determined applying the Tafel analysis, and the corrosion rate v_{corr} was calculated from the i_{corr}, according to the literature [23].

For the evaluation of the mechanism of the corrosion degradation of the ZE10 magnesium substrate with subsequent violation of the deposited Ni–P coating, the sample was immersed in the 0.1 M NaCl for 1 h. After this time, the rinsed and dried surface of the samples was analyzed via SEM.

3. Results and Discussion

3.1. Ni–P Coating Morphology and Chemical Analysis

Uniform Ni–P coating was successfully deposited on the previously pre-treated surface of the ZE10 magnesium alloy substrate. As shown in Figure 2a, grain boundaries and intermetallic phase particles appeared on the surface of the material after the pre-treated, pickled substrate. The pickled surface also shows a "honeycomb-like microstructure." Revealed intermetallic phase particles/α-Mg solid solution interface and the grain boundaries served as places where the initiation of the electroless deposition of the coating initiated due to the galvanic coupling [24]. The higher roughness of pickled and activated surface increased the adhesion of deposited Ni–P coating to the magnesium substrate.

Usually, zinc immersion is used for the magnesium alloy pre-treatment before Ni–P coating deposition. El Mahallawy et al. [14] studied the electroless Ni–P coating of different magnesium alloys, using zinc immersion as the pre-treatment of magnesium alloys. The zinc immersion was used to remove the residual oxides and hydroxides from the surface of the magnesium alloys, and a thin layer of zinc formed on the Mg surface preventing back oxidation.

Based on the obtained results, a partial re-oxidation of the ZE10 magnesium alloy surface occurred without the use of zinc immersion (Figure 2a). During the following immersion of the activated sample into the electroless nickel bath, the substrate became catalytically active when the surface oxides were dissolved in the nickel bath, and the replacement reaction occurred between the substrate and nickel ions. Even though some contamination of the substrate surface was observed before Ni–P coating deposition, it seems that it did not negatively affect the coating process [9].

The surface morphology of the deposited Ni–P coating with a nodular structure, formed by typical cauliflower-like shapes is shown in Figure 2b. Wang et al. [25] showed that deposited Ni–P coatings are formed by a columnar microstructure. However, the deposited Ni–P amorphous coatings improve the corrosion resistance of magnesium, the inherent columnar microstructure of the coating does not provide the best protection against the corrosion. The high concentration of inter-column defects, such as microvoids and micropores, [26,27], form channels where the corrosion ions and environment can pass through the coating and react with the substrate. The presence of microcavities was not evident between nodular cusps of the deposited coating (Figure 2b), which is in agreement with observations in [28]. Based on an evaluation of SEM figures (Figures 2b and 3a), no defects and cracks were observed in the deposited Ni–P coating at the ZE10 magnesium substrate/Ni–P coating interface.

The average thickness of the coating prepared for 60 min determined from the cross sections was approximately 10 µm. In the case of a longer deposition time (180 min), prepared with the aim to increase the coating thickness to obtain relevant microhardness values, thickness was 30 µm.

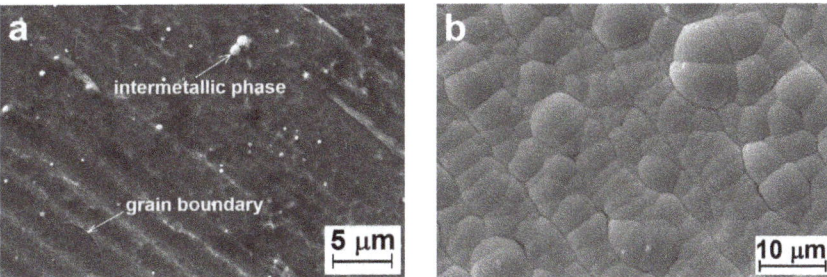

Figure 2. SEM microstructure of the pre-treated substrate and deposited Ni–P coating: (**a**) pre-treated surface of the ZE10 magnesium alloy before deposition; (**b**) fine morphology of the deposited Ni–P coating.

Deposited Ni–P coating with an average thickness of about 10 µm was chosen for EDS analysis (Figure 3). Using EDS mapping analysis, it was determined that the distribution of Ni and P in deposited Ni–P coating was homogeneous in the entire cross section, as shown in Figure 3b,c, respectively. The EDS analysis determined that the Ni content in the deposited Ni–P coating was 95.6 wt % and the P content was 4.4 wt %. Based on the literature [9,29], it was determined that the deposited Ni–P coating is low-phosphorus, as in the cases of the AZ31 magnesium alloy presented in [30] and the AZ61 magnesium alloy in [28].

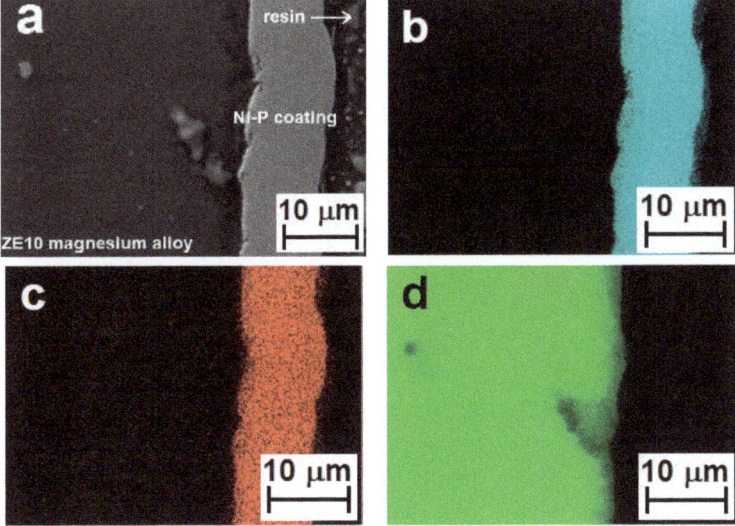

Figure 3. EDS analysis of deposited Ni–P coating on the ZE10 magnesium alloy: (**a**) a cross section of the Ni–P-coated sample, (**b**) nickel, (**c**) phosphorus, and (**d**) magnesium.

3.2. Ni–P Coating Microhardness Analysis

Based on the measured data, it was determined that the average value of the microhardness of the Ni–P coating was 690 ± 30 HV 0.025, measured in the cross section. The microhardness of the plain ZE10 magnesium substrate was 60 ± 4 HV 0.025.

The surface microhardness of the coated samples increased approximately 11-fold compared with the ZE10 magnesium alloy.

It is assumed that the measured hardness of low-phosphorus Ni–P coatings is higher compared to the high-phosphorus coatings [9]. The addition of filler (SiC, Al_2O_3) into the high-phosphorus Ni–P matrix led to a substantial increase in hardness [9]. The microhardness (690 ± 30 HV 0.025) of the deposited low-phosphorus Ni–P coating on the ZE10 magnesium alloy reached a value higher than that of the Ni–P/SiC composite coating prepared on the AZ91 magnesium alloy presented in [31] and [32]. The microhardness of the Ni–P/SiC composite coating (7.33 wt % P) was 620 HV [31], and that of the electroless Ni–P/SiC nanocomposite coating (10 wt % P) was 600 HV 0.025 [32].

The microhardness of the deposited low-phosphorus Ni–P coating was higher compared with the values obtained for the high-phosphorus Ni–P coatings. The hardness of high-phosphorus Ni–P coatings ranges from 410 to 600 HV [29,33]. As the content of phosphorus in Ni–P coatings increases, the microhardness of the coating decreased due to the microstructural changes (a decrease in crystallinity) [9].

3.3. Analysis of the Physical Properties of the Ni–P Coating

The results of the scratch test performed on the Ni–P-coated ZE10 magnesium alloy sample are shown in Figure 4. The measured values of the critical normal forces L_{c1} and L_{c2} and the corresponding friction forces F_{t1} and F_{t2}, respectively, are given in Table 2. As indicated by Table 2, the value of the critical normal force L_{c1} was 7.9 N, and the formation of oblique and parallel cracks was observed on the coating surface (Figure 5a). The value of the critical normal force L_{c2} was 13.6 N, and the formation of transverse tensile arch cracks across the entire width of the track was observed on the coating surface (Figure 5b).

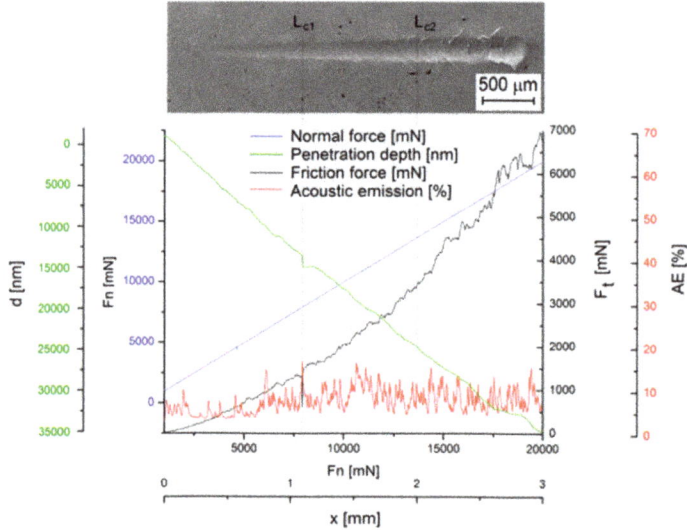

Figure 4. Results of the scratch test for the Ni–P coating on the ZE10 magnesium alloy with a scratch pattern.

As stated in [22], tensile and compressive stresses are generated during the scratch test and cause more complex mechanisms and damage. A crack can nucleate on a defect or at the coating/substrate interface. The crack is formed due to the localization of the stresses on the coating/substrate interface or in the coating (transverse crack). In the case of a layer, the tensile radial tension induced with the Rockwell tip can generate circular or transverse arch cracks that extend across the layer into the substrate. As the tip moves, several circular or transverse arch cracks can intersect. These cracks can also occur at the back of the contact as a response to tensile stresses during tip sliding. Cracks also occur on the back of the contact due to the friction-induced tensile stresses [34].

Table 2. Values of critical normal forces and friction forces of Ni–P coating deposited on the ZE10 magnesium alloy and, for comparison, values found in other studies.

Source	Substrate	Coating	L_{c1} [N]	L_{c2} [N]	F_{t1} at L_{c1} [N]	F_{t2} at L_{c2} [N]
presented Ni–P coating	ZE10	Ni–P	7.9	13.6	0.6	3.4
[30]	AZ31	Ni–P	7.3	12.3	1.1	2.6
[28]	AZ61	Ni–P	6.9	11.9	0.8	2.2
[35]	Polished AZ91	Ni–P	–	10.2 (L_c)	–	–
		Ni–P (temp.: 523 K)	–	10.6 (L_c)	–	–
		Ni–P (temp.: 673 K)	–	9.7 (L_c)	–	–
[35]	Blasted AZ91	Ni–P	–	14.0 (L_c)	–	–
		Ni–P (temp.: 523 K)	–	16.5 (L_c)	–	–
		Ni–P (temp.: 673 K)	–	14.8 (L_c)	–	–
[36]	AISI 1018	plain Ni–P/TiO$_2$	–	~13 (L_c)	–	–
		Ni–P/TiO$_2$ with SDS	–	~19 (L_c)	–	–
		Ni–P/TiO$_2$ with DTAB	–	~29 (L_c)	–	–

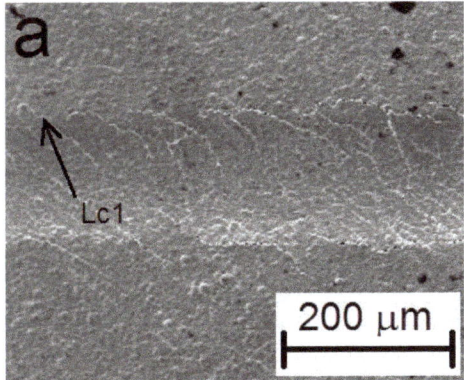 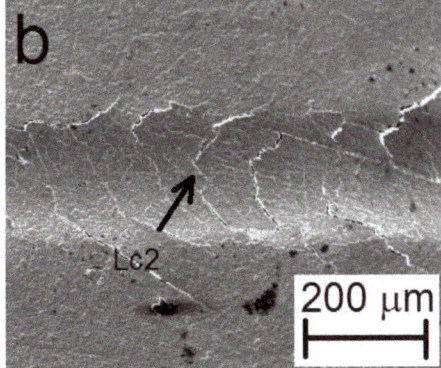

Figure 5. Detail of the damage of Ni–P coating during the scratch test: (**a**) L_{c1} and (**b**) L_{c2} (SEM).

As a result of the applied pressure load of the Rockwell diamond tip during the scratch test, ductile failure of the deposited Ni–P coating occurs due to the introduced internal stresses.

The character of the damage to the locating layer during the scratch test is dependent on many factors [22]. In addition to the influence of the characteristics of the experimental device on the tested layer damage mechanism, there are geometric properties of the substrate-layer system (such as layer thickness, roughness, etc.), experimental parameters (tip and scratch rate), and properties of the substrate-layer system (thermal coefficients, microstructure and internal stresses, elasticity, and hardness modules). Figures 4 and 5 show the scratch track morphology and the layer cracking character, which is similar to the case of Ni–P coatings on AZ31 and AZ61 magnesium alloys presented in [28,30].

The formation of transverse tensile arch cracks [22,34] across the entire width of the track was observed (Figure 4). The adhesion strength of the experimental electroless deposited Ni–P coating on a wrought ZE10 magnesium alloy (L_{c1} and L_{c2}) was higher compared to the data presented in articles [28,30], where the Ni–P coating was deposited on AZ31 and AZ61 magnesium alloys, respectively. The difference could be explained by the coated substrate pre-treatment process. The pre-treatment of AZ31 and AZ61 magnesium alloys before the deposition of the Ni–P coating included polishing to a roughness $R_a \approx 0.25$ µm [28,30]. However, the surface of the experimental ZE10 magnesium alloy was polished to a roughness $R_a \approx 2$ µm, which is significantly rougher than AZ31 and AZ61. The higher roughness of the substrate surface can improve the adhesion strength between the deposited Ni–P coating and the ZE10 magnesium alloy due to the mechanical interlocking of the two components [34].

This effect was also observed in [35], where the adhesion strength between the deposited Ni–P coating and blasted or polished surface of the AZ91 magnesium alloy was studied. The scratch track morphology for the pre-blasted and pre-polished samples with the deposited Ni–P coating showed a similar trend, but it was observed that the scratch track width was narrower on the rougher surface when compared to the polished surface. The scratch track width was slightly narrower for coated samples after annealing for 1 h at 523 K. This effect can be contributed to the increase in the hardness of the Ni–P coating after annealing, which was demonstrated with the increase in hardness from ~600 to ~900 HV due to the coated sample annealing. As indicated by Table 2, the critical load L_c for the plain Ni–P coating was 14.0 N and 10.2 N for the blasted and polished surfaces, respectively. The increase in adhesion strength to 16.5 N was observed for the rough blasted AZ91 substrate after annealing for 1 h at 523 K. This increase was apparently linked to the hardness increase and the effect of the rough surface. The brittle cracking of the deposited coating was observed at substrates with the rough surface, and the wedge spallation was observed at substrates with the polished surface. The decrease in adhesion strength was observed for samples annealed at 673 K due to the embrittlement of the Ni–P coatings (Table 2).

However, as indicated by Table 2, resulting values of the critical loads of rough (blasted) samples of AZ91 [35] are slightly higher when compared to the experimental Ni–P coating deposited on the ZE10 magnesium alloy. This can again be connected to the higher roughness of the coated substrate. The roughness of the blasted AZ91 magnesium alloy surface was $R_a \approx 4.5$ µm [35], and that of the experimental ZE10 magnesium alloy was $R_a \approx 2$ µm. It is also possible to observe that the value of the critical load L_c of experimental coating deposited on the ZE10 alloy is higher in comparison with the polished surface of the AZ91 alloy in [35], where the roughness was $R_a \approx 0.05$ µm. As is obvious, the roughness of the substrate surface has a significant effect on the coating adhesion strength due to the mechanical interlocking between Ni–P coating and the coated magnesium substrate.

As indicated in the literature [36], applied surfactants in the nickel bath had a significant effect on the adhesion strength of deposited Ni–P/TiO_2 composite coating on the AISI 1018 steel substrate (Table 2). No cohesive or adhesive failure of the coating was observed up to ~13 N in the case of the Ni–P/TiO_2 coating prepared on AISI 1018 without using the surfactant. The formation of the mild tensile cracks at ~19 N was evident for the Ni–P/TiO_2 composite coating using sodium dodecyl sulfate (SDS) surfactant at 1.5× CMC (critical micelle concentration). In the case of the Ni–P/TiO_2 composite coatings on AISI 1018 involving dodecyl trimethyl ammonium bromide (DTAB) at 1× CMC, the cohesive failure was observed at the applied load of ~29 N. Moreover, no linear or radial cracks were observed in the case of the coated steel substrate, nor of any of the analyzed coatings, which also indicates the importance of the surface of the substrate with respect to the adhesion of the coating.

The increase in the adhesion strength of Ni–P coatings to the magnesium substrate, along with a slight increase in the roughness of the substrate surface, was shown to be achieved by adding the proper surfactant into the nickel bath [35,37]. This proves that a more effective adhesion of the coatings is caused by the excessive attractive forces between the Ni–P coatings and substrate [38].

Based on the obtained result, a sufficient surface roughness of ZE10 reached via surface polishing on the roughness $R_a \approx 2$ µm, in combination with the activation of the surface via acid pickling, seems to be reached during pre-treatment. Adequate pretreatment resulted in an adequate adhesion of the coating to the substrate and a considerably high resistivity against damage.

3.4. The Electrochemical Corrosion Test in 0.1 M NaCl

Figure 6 shows the potentiodynamic polarization curves of the ZE10 magnesium alloy and the ZE10 alloy with the deposited Ni–P coating in 0.1 M NaCl obtained at laboratory temperature. The polarization curve of the Ni–P-coated sample is significantly shifted to more electropositive values, which means better corrosion properties of the Ni–P-coated sample compared with the untreated ZE10 magnesium alloy.

Therefore, a deposited Ni–P coating appears to be suitable for the protection of magnesium alloys. Based on the Tafel extrapolation analysis [39], the values of the corrosion potential, E_{corr}, and the corrosion current density, i_{corr}, for samples with the deposited Ni–P coating and the ZE10 magnesium alloy were determined. The average values of the E_{corr} for the ZE10 magnesium alloy and the Ni–P-coated alloy, were −1701 and −505 mV, respectively, and the values of the i_{corr} for the ZE10 magnesium alloy and the Ni–P-coated material, were 23.7 and 0.4 µA·cm^{-2}, respectively (Table 3). It is generally known [25] that the columnar structure of deposited Ni–P coatings contains a network of defects (grain boundaries, pores, and microcavities). These defects are precursors for a micropitting and may cause the corrosion attack of the material under the deposited Ni–P coating. However, no local corrosion attack (pitting) was observed in the anodic area of the polarization curves.

Based on the determined values of i_{corr}, the corrosion rate, v_{corr}, was calculated for a short-term experiment. As indicated by Table 3, the corrosion rates of the ZE10 alloy and the Ni–P-coated samples in 0.1 M NaCl were 530.00 µmpy and 8.95 µmpy, respectively.

Experimental deposited Ni–P coating was ranked among the low-phosphorus Ni–P coatings, which are characterized by their lower corrosion resistance when compared to the medium- and high-phosphorus coatings, [9]. As indicated in the work of S. Narayanan [12], deposited Ni–P coatings with low phosphorus content (3.34 wt % P) showed an E_{corr} of −0.536 V and an i_{corr} of 4.22 µA·cm^{-2}, whereas medium- (6.70 wt % P) and high-phosphorus (13.30 wt % P) Ni–P coatings showed E_{corr} values of −0.434 and −0.411 V, respectively, and i_{corr} values of 1.17 and 0.60 µA·cm^{-2}, respectively. When comparing the experimentally obtained results and the results presented in [12], it is evident that the value (Table 3) of the corrosion potential, E_{corr}, is between E_{corr} values of 25-µm-thick low-phosphorus (3.34 wt % P) and medium-phosphorus (6.70 wt % P) Ni–P coatings presented in [12].

Table 3. Corrosion parameters of the ZE10 magnesium alloy and the ZE10 alloy with a deposited Ni–P coating.

Sample	E_{corr} [mV]	i_{corr} [µA·cm^{-2}]	v_{corr} [µmpy]
ZE10 alloy	−1701	23.7	530.00
ZE10 + Ni–P coating	−505	0.4	8.95

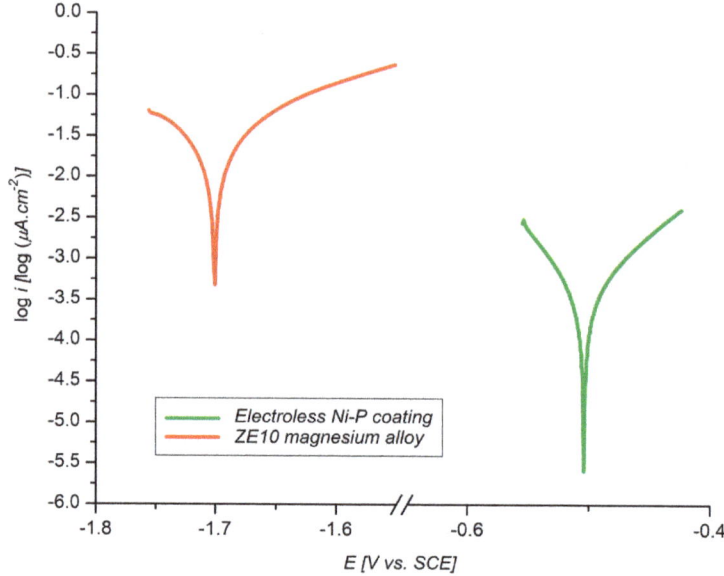

Figure 6. Characteristic potentiodynamic polarization curves for the ZE10 magnesium alloy and the ZE10 alloy with a deposited Ni–P coating in 0.1 M NaCl obtained at laboratory temperature.

Although the plating rate of high-phosphorus Ni–P coatings in [12] was higher, high-phosphorus Ni–P coatings were deposited under more energy-intensive conditions (90 ± 1 °C) [12]. It is evident that the obtained value of the corrosion current density, i_{corr}, of high-phosphorus Ni–P coating was worse compared to the presented low-phosphorus Ni–P coatings on the ZE10 magnesium alloy (Table 3), and the electroless deposited coating on the ZE10 alloy exhibits greater corrosion resistance. This can be attributed to the fact that the experimentally prepared electroless deposited Ni–P coating was probably less defective or that the Ni–P coating contained only a small amount of microcavities when compared to the coatings analyzed in [12].

3.5. The Immersion Test in 0.1 M NaCl

As indicated by Figure 7a,b, after the exposition of the sample with deposited Ni–P coating in 0.1 M NaCl for 1 h, the degradation of the Ni–P coating occurred due to the corrosion of the magnesium substrate under the coating.

Corrosive agents may accumulate between the nodules (grains) creating the Ni–P coating and migrate to the substrate surface. According to the results in [25,40], the inherent columnar porous microstructure of the coating (Figure 7d) does not adequately protect the magnesium substrate against corrosive environments [41]. Furthermore, electroless Ni–P coatings contain a certain amount of microcavities in their volume. These microcavities form nucleation sites for micropitting when the material is exposed to a corrosive environment. Microcavities are created as a result of the hydrogen evolution during the deposition process when small bubbles of hydrogen H_2, as a byproduct of the nickel reduction, are adsorbed on the surface of the growing Ni–P coating [40]. Stirring of the nickel baths or the rotation of the deposited objects is a partial solution of this problem. However, the addition of surfactants into the nickel bath was shown to be a more effective solution, while the reduction of these microcavities resulted in an increase in the corrosion resistance of the coated substrate [42,43].

As described above, due to the high concentration of inter-column defects, such as microvoids and micropores [27], the interaction between the corrosive environment containing chloride ions and magnesium substrate under the deposited Ni–P coating resulted in the creation of oxides or chlorides of magnesium. This reaction has an effect on the increase in corrosion product volume when compared to the bulk material and the evolution of hydrogen is an accompanying process. The increase in the volume of the material under the Ni–P coating leads to the formation of cracks and the subsequent destruction of the coating and thus to the acceleration of the corrosion of the magnesium substrate (Figure 7c).

Even though the coating internal defects were not observed in SEM analysis, even the intercolumnar areas can provide a path for the transfer of corrosive medium elements into the coating and react with the substrate. Figure 7d reveals the damaged structure of the Ni–P coating showing the columnar structure. The revealed columnar structure supports the theory that the grain boundaries and microdefects present on the grain boundaries acted as the paths for the corrosive medium transfer to the substrate.

Observed local corrosion attack did not correlate with the results obtained with potentiodynamic tests; however, the electrochemical test was limited to the range of polarization used, and a larger range for the measurement likely can reveal the pitting attack.

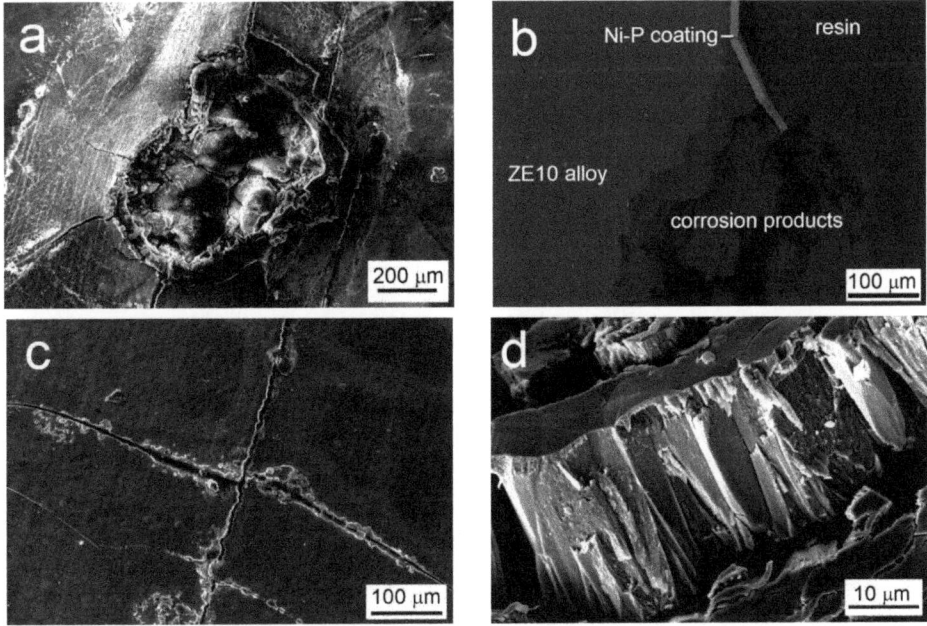

Figure 7. Corrosion degradation of the Ni–P coating on the ZE10 magnesium alloy: (**a**) surface, (**b**) cross-section, (**c**) cracks in Ni–P coating caused by the increase in volume under deposited coating due to the corrosion, (**d**) cracked Ni–P coating.

4. Conclusions

Based on the results of the characterization of the electroless deposited Ni–P coating prepared on a ZE10 magnesium alloy via polishing and acid pickling, the following conclusions can be made:

- Due to the proper pre-treatment, electroless low-phosphorus Ni–P coatings were successfully deposited on a ZE10 magnesium alloy in a nickel bath.
- The content of Ni in the deposited coatings was 95.6 wt %, and the content of P was 4.4 wt %, for coatings with average thickness ~10 µm (1 h of deposition).
- Deposited Ni–P coatings showed a high degree of homogeneity in the entire cross section.

- The microhardness of the deposited low-phosphorus Ni–P coating was 690 ± 30 HV 0.025, representing an 11-fold increase when compared to the plain ZE10 magnesium alloy.
- The adhesion of the deposited Ni–P coating was determined from the appropriate critical normal forces L_{c1} = 7.9 N, where i.e., oblique and parallel cracks are the primary failure—was observed, and L_{c2} = 13.6 N, where the formation of transverse arch cracks was observed.
- Electrochemical polarization tests have shown an improvement in corrosion resistance via the deposition of Ni–P coating (E_{corr} = −505 mV, i_{corr} = 0.4 µA·cm^{-2}, and v_{corr} = 8.95 µmpy) when compared to the ZE10 magnesium alloy (E_{corr} = −1701 mV, i_{corr} = 23.7 µA·cm^{-2}, and v_{corr} = 530 µmpy).
- Microcavities present in the coating and the columnar structure of the low-phosphorus Ni–P coating have a major influence on the process of the corrosion, acting as the channels between the corrosive environment and the substrate.

Acknowledgments: This work was supported by project No LO1211, Materials Research Centreat FCH BUT-Sustainability and Development (National Programme for Sustainability I, Ministry of Education, Youth and Sports).

Author Contributions: Martin Buchtík designed and performed experiments; Petr Kosár prepared samples and performed experiments; Jakub Tkacz analyzed data and performed electrochemical measurement; Jaromír Wasserbauer contributed materials and analysis tools, and analyzed data; Pavel Doležal performed adhesion tests and analyzed data. Martin Buchtík wrote the paper.

Conflicts of Interest: The authors declare no conflict of interest.

References

1. Friedrich, H. Alloys Containing Zirconium. In *Magnesium Technology Metallurgy, Design Data, Automotive Applications*; Mordike, B.L., Ed.; Springer: Berlin, Germany, 2004; pp. 140–141.
2. ASM International Handbook Committee. *Properties and Selection: Nonferrous Alloys and Special-Purpose Materials*, 10th ed.; ASM International: Materials Park, OH, USA, 1990.
3. Miao, X.; Li, X.; Hu, H.; Gao, G.; Zhang, S. Effects of the Oxide Coating Thickness on the Small Flaw Sizing Using an Ultrasonic Test Technique. *Coatings* **2018**, *8*, 69. [CrossRef]
4. Shao, Z.; Cai, Z.; Hu, R.; Wei, S. The study of electroless nickel plating directly on magnesium alloy. *Surf. Coat. Technol.* **2014**, *249*, 42–47. [CrossRef]
5. Liu, Y.; Li, W.; Li, Y. Microstructure and mechanical properties of ZE10 magnesium alloy prepared by equal channel angular pressing. *Int. J. Min. Met. Mater.* **2009**, *16*, 559–563. [CrossRef]
6. Bettles, C.; Barnett, M. *Advances in Wrought Magnesium Alloys: Fundamentals of Processing, Properties and Applications*; Woodhead Publishing: Oxford, UK, 2012.
7. Fauchais, P.L.; Heberlein, J.V.R.; Boulos, M.I. *Thermal Spray Fundamentals: From Powder to Part*; Springer: New York, NY, USA, 2013.
8. Drábiková, J.; Fintová, S.; Tkacz, J.; Doležal, P.; Wasserbauer, J. Unconventional fluoride conversion coating preparation and characterization. *Anti-Corros. Method Mater.* **2017**, *64*, 613–619. [CrossRef]
9. Riedel, W. *Electroless Nickel Plating*; ASM International: Metals Park, OH, USA, 1991.
10. Mallory, G.O.; Hajdu, J.B. *Electroless Plating: Fundamentals and Applications*; Knoyes Publications/William Andrew Publishing: Norwich, NY, USA, 2009.
11. Abrantes, L.M.; Fundo, A.; Jin, G. Influence of phosphorus content on the structure of nickel electroless deposits. *J. Mater. Chem.* **2001**, *11*, 200–203. [CrossRef]
12. Sankara Narayanan, T.S.N.; Baskaran, I.; Krishnaveni, K.; Parthiban, S. Deposition of electroless Ni–P graded coatings and evaluation of their corrosion resistance. *Surf. Coat. Technol.* **2006**, *200*, 3438–3445. [CrossRef]
13. Curtis, J.M.; Polak, T.A.; Wilcox, G.D. *Handbook of Surface Treatments and Coatings*; Professional Engineering Publishing: London, UK, 2003.
14. El Mahallawy, N.; Bakkar, A.; Shoeib, M.; Palkowski, H.; Neubert, V. Electroless Ni–P coating of different magnesium alloys. *Surf. Coat. Technol.* **2008**, *202*, 5151–5157. [CrossRef]
15. Wu, L.; Zhao, J.; Xie, Y.; Yang, Z. Progress of electroplating and electroless plating on magnesium alloy. *Trans. Nonferr. Metal. Soc.* **2010**, *20*, 630–637. [CrossRef]

16. Zhou, Y.; Zhang, S.; Nie, L.; Zhu, Z.; Zhang, J.; Cao, F.; Zhang, J. Electrodeposition and corrosion resistance of Ni–P–TiN composite coating on AZ91D magnesium alloy. *Trans. Nonferr. Metal. Soc. Chin.* **2016**, *26*, 2976–2987. [CrossRef]
17. Chen, J.; Yu, G.; Hu, B.; Liu, Z.; Ye, L.; Wang, Z. A zinc transition layer in electroless nickel plating. *Surf. Coat. Technol.* **2006**, *201*, 686–690. [CrossRef]
18. *ASTM B90/B90M-15 Standard Specification for Magnesium-Alloy Sheet and Plate*; ASTM International: West Conshohocken, PA, USA, 2015.
19. Vander Voort, G.F. *Metallography, Principles and Practice*; ASM International: Materials Park, OH, USA, 1999.
20. Tran, T.N.; Yu, G.; Hu, B.N.; Xie, Z.H.; Tang, R.; Zhang, X.Y. Effects of pretreatments of magnesium alloys on direct electroless nickel plating. *Trans. IMF* **2012**, *90*, 209–214. [CrossRef]
21. *ASTM E384-17 Standard Test Method for Microindentation Hardness of Materials*; ASTM International: West Conshohocken, PA, USA, 2017.
22. *ASTM C1624-05(2015) Standard Test Method for Adhesion Strength and Mechanical Failure Modes of Ceramic Coatings by Quantitative Single Point Scratch Testing*; ASTM International: West Conshohocken, PA, USA, 2015.
23. Szewieczek, D.; Baron, A. Electrochemical corrosion and its influence on magnetic properties of $Fe_{75.5}Si_{13.5}B_9Nb_3Cu_1$ alloy. *J. Mater. Process. Technol.* **2005**, *164–165*, 940–946. [CrossRef]
24. Ambat, R.; Zhou, W. Electroless nickel-plating on AZ91D magnesium alloy: effect of substrate microstructure and plating parameters. *Surf. Coat. Technol.* **2004**, *179*, 124–134. [CrossRef]
25. Wang, W.; Zhang, W.; Wang, Y.; Mitsuzak, N.; Chen, Z. Ductile electroless Ni–P coating onto flexible printed circuit board. *Appl. Surf. Sci.* **2016**, *367*, 528–532. [CrossRef]
26. Yang, H.; Gao, Y.; Qin, W. Investigation of the Corrosion Behavior of Electroless Ni–P Coating in Flue Gas Condensate. *Coatings* **2017**, *7*, 16. [CrossRef]
27. Mao, S.; Yang, H.; Li, J.; Huang, F.; Song, Z. Corrosion properties of aluminium coatings deposited on sintered NdFeB by ion-beam-assisted deposition. *Appl. Surf. Sci.* **2011**, *257*, 5581–5585. [CrossRef]
28. Buchtík, M.; Kosár, P.; Wasserbauer, J.; Doležal, P. Characterization of Ni–P coating prepared on a wrought AZ61 magnesium alloy via electroless deposition. *Mater. Tehnol.* **2017**, *51*, 925–931. [CrossRef]
29. Parkinson, R. *Properties and Applications of Electroless Nickel*; Technical paper 10081; Nickel Development Institute: Toronto, ON, Canada, 1997.
30. Buchtík, M.; Kosár, P.; Wasserbauer, J.; Doležal, P. Characterization of Ni–P coating prepared via electroless deposition on AZ31 magnesium alloy. *Koroze a ochrana materiálu* **2017**, *61*, 1–6. [CrossRef]
31. Wang, H.-L.; Liu, L.-Y.; Dou, Y.; Zhang, W.-Z.; Jiang, W.-F. Preparation and corrosion resistance of electroless Ni–P/SiC functionally gradient coatings on AZ91D magnesium alloy. *Appl. Surf. Sci.* **2013**, *286*, 319–327. [CrossRef]
32. Farzaneh, A.; Mohammadi, M.; Ehteshamzadeh, M.; Mohammadi, F. Electrochemical and structural properties of electroless Ni-P-SiC nanocomposite coatings. *Appl. Surf. Sci.* **2013**, *276*, 697–704. [CrossRef]
33. Balaraju, J.N.; Sankara Narayanan, T.S.N.; Seshadri, S.K. Electroless Ni–P composite coatings. *J. Appl. Electrochem.* **2003**, *33*, 807–816. [CrossRef]
34. Holmberg, K.; Matthews, A. *Coatings Tribology Properties, Mechanisms, Techniques and Applications in Surface Engineering*, 2nd ed.; Elsevier Science: Amsterdam, The Netherlands, 2009.
35. Liu, Z.; Gao, W. Electroless nickel plating on AZ91 Mg alloy substrate. *Surf. Coat. Technol.* **2006**, *200*, 5087–5093. [CrossRef]
36. Tamilarasan, T.R.; Rajendran, R.; Siva shankar, M.; Sanjith, U.; Rajagopal, G.; Sudagar, J. Wear and scratch behaviour of electroless Ni–P-nano-TiO_2: Effect of surfactants. *Wear* **2016**, *346–347*, 148–157. [CrossRef]
37. Elansezhian, R.; Ramamoorthy, B.; Kesavan Nair, P. Effect of surfactants on the mechanical properties of electroless (Ni–P) coating. *Surf. Coat. Technol.* **2008**, *203*, 709–712. [CrossRef]
38. Tracton, A.A. *Coatings Technology: Fundamentals, Testing, and Processing Techniques*; CRC Press: Boca Raton, FL, USA, 2007.
39. Tkacz, J.; Minda, J.; Fintová, S.; Wasserbauer, J. Comparison of Electrochemical Methods for the Evaluation of Cast AZ91 Magnesium Alloy. *Materials* **2016**, *9*, 925. [CrossRef] [PubMed]
40. Chen, B.-H.; Hong, L.; Ma, Y.; Ko, T.-M. Effects of Surfactants in an Electroless Nickel-Plating Bath on the Properties of Ni−P Alloy Deposits. *Ind. Eng. Chem. Res.* **2002**, *41*, 2668–2678. [CrossRef]

41. Valova, E.; Georgieva, J.; Armyanov, S.; Avramova, I.; Dille, J.; Kubova, O.; Delplancke-Ogletree, M.-P. Corrosion behavior of hybrid coatings: Electroless Ni–Cu–P and sputtered TiN. *Surf. Coat. Technol.* **2010**, *204*, 2775–2781. [CrossRef]
42. Muraliraja, R.; Elansezhian, R.; Sudagar, J.; Raviprakash, A.V. Influence of a Zwitterionic Surfactant on the Surface Properties of Electroless Ni–P Coating on Mild Steel. *J. Surfactants Deterg.* **2016**, *19*, 1081–1088. [CrossRef]
43. Latt, K.M. Effects of Surfactants on Characteristics and Applications of Electroless Nickel-Phosphorous Deposits. Master's Thesis, National University of Singapore, Singapore, 2003.

© 2018 by the authors. Licensee MDPI, Basel, Switzerland. This article is an open access article distributed under the terms and conditions of the Creative Commons Attribution (CC BY) license (http://creativecommons.org/licenses/by/4.0/).

Article

Internally Oxidized Ru–Zr Multilayer Coatings

Yung-I Chen *, Tso-Shen Lu and Zhi-Ting Zheng

Institute of Materials Engineering, National Taiwan Ocean University, 2 Pei-Ning Road, Keelung 20224, Taiwan; x76825@gmail.com (T.-S.L.); 10455001@ntou.edu.tw (Z.-T.Z.)
* Correspondence: yichen@mail.ntou.edu.tw; Tel.: +886-2-2462-2192

Academic Editors: Tony Hughes and Russel Varley
Received: 20 February 2017; Accepted: 21 March 2017; Published: 23 March 2017

Abstract: In this study, equiatomic Ru–Zr coatings were deposited on Si wafers at 400 °C by using direct current magnetron cosputtering. The plasma focused on the circular track of the substrate holder and the substrate holder rotated at speeds within 1–30 rpm, resulting in cyclical gradient concentration in the growth direction. The nanoindentation hardness levels of the as-deposited Ru–Zr coatings increased as the stacking periods of the cyclical gradient concentration decreased. After the coatings were annealed in a 1% O_2–99% Ar atmosphere at 600 °C for 30 min, the internally oxidized coatings shifted their respective structures to a laminated structure, misaligned laminated structure, and nanocomposite, depending on their stacking periods. The effects of the stacking period of the cyclical gradient concentration on the mechanical properties and structural evolution of the annealed Ru–Zr coatings were investigated in this study.

Keywords: cyclical gradient concentration; internal oxidation; multilayer coating; nanocomposite coating

1. Introduction

Multilayer nitride coatings with nanoscale layer thickness have exhibited extremely high mechanical hardness due to dislocation blocking by layer interfaces and Hall–Petch strengthening [1]. By contrast, the hardness enhancement in the Y_2O_3/ZrO_2 superlattice has been limited because oxides are brittle materials that are deformed by fracture mechanisms [2]. Two metallic multilayer coatings deposited by cosputtering for immiscible systems, W–Cu [3,4] and Cu–Ta [5,6], have developed a phase-separated nanostructure. However, Ru/Al multilayers have been deposited to fabricate a B2-RuAl intermetallic compound through annealing at approximately 600 °C in a vacuum or Ar [7,8]. Oxide-dispersion-strengthened platinum materials [9] and Ag-oxide-based electric contact material [10] are conventional applications of internal oxidation [11]. Our previous studies [12–15] investigated the internal oxidation of Ru-based alloy multilayer coatings annealed at 600 °C in oxygen-containing atmospheres for the application of protective coatings on glass molding dies. The specific cosputtering processes, which were performed using a substrate holder rotating at a slow speed of one to seven revolutions per minute, have been examined in detail for fabricating Ru–Ta coatings [14]; the fabricated coatings had exposed substrates alternately to the sputter sources without shutter shielding, forming a multilayer structure with a cyclical gradient concentration period at a nanometer scale. An oxidized laminated structure formed because of the inward diffusion of oxygen during the annealing process; this structure comprised alternating oxygen-rich and oxygen-deficient sublayers stacked adjacent to the surface. The inward diffusion of oxygen at 600 °C was dominated by lattice diffusion in the active element-enriched regions [13,16,17]. Because the elements were stacked on the substrate with an alternating gradient concentration, the O atoms could easily diffuse through the paths in the transverse direction, thereby forming oxide sublayers. After the oxygen content in the oxide sublayers reached a saturation level, the grainboundary diffusion along the original columnar structure drove oxygen to the next period of the laminated structure. During an annealing process conducted at 600 °C in a 1% O_2–99% Ar atmosphere, internal oxidation occurred for Ti–Ru, Zr–Ru, Nb–Ru, Mo–Ru, Hf–Ru, Ta–Ru, and W–Ru coatings, which were prepared using a substrate holder rotating at one revolution per minute [15]; the mechanical properties of the annealed coatings depended on the characteristics of the oxide sublayers. The nanoindentation hardness of the annealed $Zr_{0.30}Ru_{0.70}$ coating exhibited a relatively high value of 18.4 GPa. The widths of the oxide sublayers were restricted by the Ru-dominant sublayers [16,17]; therefore, the internally oxidized coatings can be categorized as nonisostructural oxide/metal multilayers [1]. The substrate holder rotation speed in sputtering affects the stacking period of the laminated structure [14]; therefore, assessing the effect of the stacking period on the mechanical properties of the internally oxidized Ru–Zr coatings is imperative.

2. Materials and Methods

Ru–Zr coatings with a Cr interlayer were fabricated by using magnetron cosputtering onto silicon wafers. Pure metal targets of Ru (99.95%), Zr (99.9%), and Cr (99.95%) with diameters of 50.8 mm each were adopted as source materials for sputtering. The sputter guns were inclined to focus plasma on the circular track of the substrate holder, as described in detail in a previous study [13]. The target-to-substrate distance was maintained at 90 mm for all sputtering runs. The chamber was evacuated down to 2.7×10^{-4} Pa, followed by the inlet of argon gas as a plasma source. The substrate holder was heated to 400 °C and the Ar flow rate was controlled at 20 sccm; the resulting working pressure was 0.7 Pa. The substrate holder was rotated at 1 rpm for depositing the Cr interlayer. Then, Ru–Zr coatings with fixed DC sputtering powers (W_{Ru} = 100 W and W_{Zr} = 200 W) and various substrate holder rotation speeds were deposited on the Cr interlayer for 25 min. To investigate the internal oxidation phenomenon after performing heat treatments, the Ru–Zr coatings were further annealed at 600 °C in a 1% O_2–99% Ar atmosphere by introducing O_2–Ar mixed gas into a quartz tube furnace.

Chemical composition analysis was conducted by using energy dispersive spectrometry (EDS, Horiba, Kyoto, Japan) equipped with a scanning electron microscope (SEM, S3400N, Hitachi, Tokyo, Japan) on the surface. Surface morphology and thickness evaluation of the coatings were performed by using a field emission scanning electron microscope (FE-SEM, S4800, Hitachi, Tokyo, Japan) at a 15-kV accelerating voltage. A conventional X-ray diffractometer (XRD, X'Pert PRO MPD, PANalytical, Almelo, The Netherlands) with Cu Kα radiation was adopted to identify the phases of the coatings, using the grazing incidence technique with an incidence angle of 1°. The Cu Kα radiation was generated from a Cu anode operated at 45 KV and 40 mA. The nanostructure was examined by using a transmission electron microscope (TEM, JEM-2010F, JEOL, Tokyo, Japan) at a 200-kV accelerating voltage. The TEM samples were prepared by applying a focused ion beam system (FEI Nova 200, Hillsboro, OR, USA) operated at an accelerating voltage of 30 kV with a gallium ion source. A Pt layer was deposited to protect the free surface in the sample preparation. The chemical states of the constituent elements were examined by using an X-ray photoelectron spectroscope (XPS, PHI 1600, PHI, Kanagawa, Japan) with an Mg Kα X-ray beam (energy = 1253.6 eV and power = 250 W) operated at 15 kV. The XPS spectra of O 1s, Ru 3d, and Zr 3d core levels were recorded. Ar+ ion beam of 3 keV was used to sputter the coatings for depth profiling. The surface hardness and Young's modulus of Ru–Zr coatings were measured with a nanoindentation tester (TI-900 Triboindenter, Hysitron, Minneapolis, MN, USA). The nanoindenter was equipped with a Berkovich diamond-probe tip. The applied load was controlled to produce an indentation depth of 80 nm, which is 1/10 of the film thickness [18]. The loading, holding, and unloading times were 5 s each. The nanoindentation hardness and elastic modulus of each indent were calculated using the Oliver and Pharr method [19]. The standard deviations for nanoindentation data were calculated from 8 measurements made at different locations on one sample. The surface roughness values of the coatings, R_a [20], were evaluated by using an atomic force microscope (AFM, Dimension 3100 SPM, NanoScope IIIa, Veeco, Santa Barbara, CA, USA). The scanning area of each image was set at 5×5 μm^2 with a scanning rate of 1.0 Hz.

3. Results

3.1. As-Deposited Equiatomic Ru–Zr Coatings

Table 1 lists the chemical compositions of the as-deposited equiatomic Ru–Zr coatings prepared at various substrate holder rotation speeds of 1–30 rpm. The samples were denoted as $Ru_xZr_{1-x}(Ry)$, or Ry, where Ry indicated that the sample prepared using the substrate holder was rotated at y rpm. All the coatings exhibited similar atomic ratios Ru/(Ru + Zr) within 0.46–0.50 after being examined using EDS on the surface, and a thickness of 870–920 nm after being evaluated using FE-SEM in the cross section. Oxygen content in the as-deposited coatings was 0.1–0.5 at.% because of weak oxidation caused by the residual oxygen in the vacuum chamber.

Table 1. Chemical compositions, thickness values, laminated period, mechanical properties, and surface roughness values of $Ru_xZr_{1-x}(Ry)$ coatings as-deposited and annealed at 600 °C in 1% O_2–99% Ar for 30 min.

Sample	Chemical Composition (at.%)			Atomic Ratio	Thickness (nm)		Period (nm)	H (GPa)	E (GPa)	Roughness (nm)
	Ru	Zr	O	Ru:Zr	Coating	Interlayer				
As-deposited										
$Ru_{0.50}Zr_{0.50}$(R1)	50.2	49.6	0.2	50.3:49.7	870	90	34.8	9.1 ± 0.2	128 ± 3	1.76 ± 0.04
$Ru_{0.49}Zr_{0.51}$(R3)	48.7	50.9	0.4	48.9:51.1	900	100	12.0	10.3 ± 0.3	142 ± 3	2.51 ± 0.02
$Ru_{0.48}Zr_{0.52}$(R5)	47.9	51.8	0.3	48.0:52.0	890	100	7.2	10.5 ± 0.6	137 ± 6	3.35 ± 0.06
$Ru_{0.47}Zr_{0.53}$(R10)	47.2	52.3	0.5	47.4:52.6	900	100	3.6	11.0 ± 0.4	161 ± 4	2.62 ± 0.05
$Ru_{0.46}Zr_{0.54}$(R15)	46.3	53.4	0.3	46.4:53.6	900	90	2.4	11.1 ± 0.5	177 ± 7	4.08 ± 0.04
$Ru_{0.47}Zr_{0.53}$(R20)	46.6	53.4	0.1	46.6:53.4	920	95	1.8	11.4 ± 0.6	171 ± 6	1.25 ± 0.01
$Ru_{0.46}Zr_{0.54}$(R30)	46.4	53.6	0.1	46.4:53.6	920	90	1.2	13.1 ± 0.5	172 ± 5	1.37 ± 0.01
Annealed										
$Ru_{0.50}Zr_{0.50}$(R1)	21.3	21.1	57.5	50.1:49.9	1380	110	55.2	15.5 ± 0.4	157 ± 10	5.33 ± 0.50
$Ru_{0.49}Zr_{0.51}$(R3)	21.2	21.2	59.2	48.0:52.0	1370	110	18.1	16.1 ± 0.8	158 ± 8	4.26 ± 0.10
$Ru_{0.48}Zr_{0.52}$(R5)	20.6	20.6	60.7	47.6:52.4	1390	110	11.1	17.2 ± 0.4	178 ± 9	7.02 ± 0.37
$Ru_{0.47}Zr_{0.53}$(R10)	20.3	20.3	61.5	47.3:52.7	1390	110	5.5	12.3 ± 2.1	164 ± 16	17.32 ± 0.53
$Ru_{0.46}Zr_{0.54}$(R15)	20.2	20.2	62.5	46.1:53.9	1390	110	3.7	16.4 ± 1.0	182 ± 6	7.05 ± 0.20
$Ru_{0.47}Zr_{0.53}$(R20)	20.8	20.8	61.3	46.3:53.7	1380	110	2.8	16.1 ± 0.8	160 ± 6	1.89 ± 0.00
$Ru_{0.46}Zr_{0.54}$(R30)	21.0	21.0	61.4	45.7:54.3	1390	110	1.9	17.9 ± 0.7	175 ± 6	5.90 ± 1.05

Figure 1 shows cross-sectional SEM images of the as-deposited Ru–Zr coatings, which exhibit a columnar structure. Laminated structures stacked along the growth direction were observed in the $Ru_{0.50}Zr_{0.50}$(R1) and $Ru_{0.49}Zr_{0.51}$(R3) coatings, for which the equilibrated laminated layer periods were 35 and 12 nm, respectively, as determined using the thickness recorded from the SEM observation divided by the number of laminated layers; in other words, the number of revolutions of the substrate holder. Each equilibrated laminated layer period formed as a result of cyclical gradient concentration deposition. The laminated structures of the Ru–Zr(Ry) coatings prepared at higher substrate holder rotation speeds such as $Ru_{0.47}Zr_{0.53}$(R10) and $Ru_{0.46}Zr_{0.54}$(R30) exhibited narrower equilibrated laminated layer periods that could not be evaluated through SEM images.

Figure 2 shows the XRD patterns of the as-deposited Ru–Zr(Ry) coatings. The $Ru_{0.50}Zr_{0.50}$(R1), $Ru_{0.49}Zr_{0.51}$(R3), $Ru_{0.48}Zr_{0.52}$(R5), and $Ru_{0.47}Zr_{0.53}$(R10) coatings exhibited reflections of hexagonal Ru [ICDD 06-0663], cubic RuZr [ICDD 18-1147], and hexagonal Zr [ICDD 05-0665] phases, implying that these coatings consisted of laminated sublayers. The equilibrated laminated layer periods for the R5 and R10 coatings were 7.2 and 3.6 nm, respectively. By contrast, XRD patterns of the as-deposited $Ru_{0.46}Zr_{0.54}$(R15), $Ru_{0.47}Zr_{0.53}$(R20), and $Ru_{0.46}Zr_{0.54}$(R30) coatings exhibited a RuZr phase dominant structure. The cubic RuZr phase exhibited XRD reflections of (110), (200), and (211), which are comparable with previous XRD results reported by Mahdouk et al. [21]. RuZr exhibited a B2 structure (CsCl type) [21–25]. Figure 3 depicts a cross-sectional TEM image of the as-deposited $Ru_{0.46}Zr_{0.54}$(R15) coating, which comprises a columnar structure without evident laminated sublayers; the diffraction pattern of the selected area shows a cubic RuZr phase. The equilibrated laminated layer periods for the as-deposited $Ru_{0.46}Zr_{0.54}$(R15), $Ru_{0.47}Zr_{0.53}$(R20), and $Ru_{0.46}Zr_{0.54}$(R30) coatings were 2.4, 1.8, and 1.2 nm, respectively, which were too thin to construct the laminated structure. Under such conditions, the equilibrated laminated layer periods were equal to a variation period of cyclical gradient concentration. Because the substrate temperature was sustained at 400 °C during cosputtering, the deposited atoms formed an intermetallic RuZr compound, as observed by the XRD patterns. In our previous study [26], B2-RuAl phase was observed for Ru–Al multilayer coatings prepared at 400 °C.

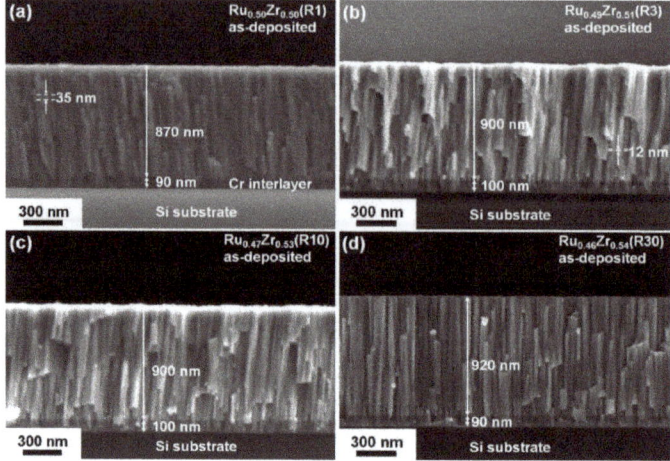

Figure 1. Cross-sectional SEM images of the as-deposited (**a**) $Ru_{0.50}Zr_{0.50}$(R1); (**b**) $Ru_{0.49}Zr_{0.51}$(R3); (**c**) $Ru_{0.47}Zr_{0.53}$(R10); and (**d**) $Ru_{0.46}Zr_{0.54}$(R30) coatings.

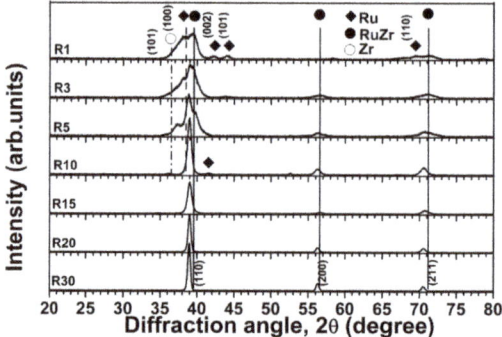

Figure 2. XRD patterns of the as-deposited Ru–Zr(Ry) coatings.

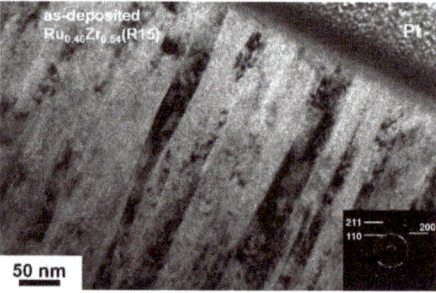

Figure 3. Cross-sectional TEM image and selected area diffraction pattern of the as-deposited $Ru_{0.46}Zr_{0.54}$(R15) coating.

3.2. Internally Oxidized Ru–Zr Coatings

Figure 4 shows the cross-sectional SEM image of the annealed $Ru_{0.50}Zr_{0.50}$(R1) coating, which exhibited a laminated structure with an equilibrated laminated layer period of 55 nm. However, the features of the other coatings could not be identified through SEM. Figure 5 presents the XRD patterns of the Ru–Zr coatings after annealing in 1% O_2–99% Ar at 600 °C for 30 min; all patterns exhibited monoclinic ZrO_2 [ICDD 32-1484], tetragonal ZrO_2 [ICDD 42-1164], and Ru phases. The Ru:Zr atomic ratios were maintained at levels similar to those of the as-deposited coatings (Table 1), implying that no volatile oxides were formed during annealing. The O content in the annealed coatings increased to within 58–62 at.%, indicating that extra O was trapped because the stoichiometric ratio of ZrO_2 was two, enabling partial Ru atoms to be oxidized.

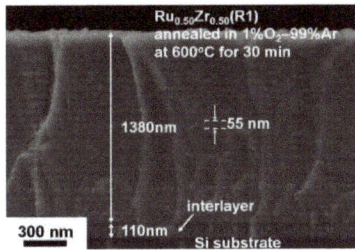

Figure 4. Cross-sectional SEM image of the $Ru_{0.50}Zr_{0.50}$(R1) coating after annealing in 1% O_2–99% Ar at 600 °C for 30 min.

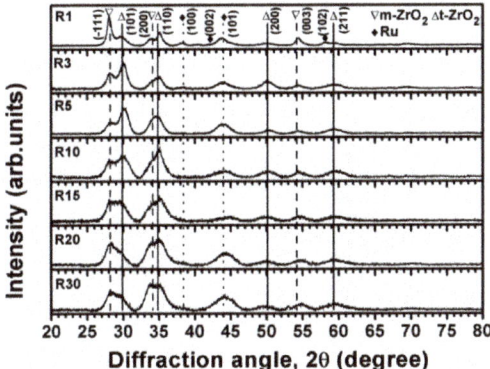

Figure 5. XRD patterns of the Ru–Zr(Ry) coatings after annealing in 1% O_2–99% Ar at 600 °C for 30 min.

Figure 6a–c illustrates the XPS spectra of O 1s, Zr 3d, and Ru 3d core levels, respectively, at various thickness levels of the annealed $Ru_{0.50}Zr_{0.50}$(R1) coating. The detected depth crossed six periods of the laminated layers. The O and Zr species were identified as O^{2-} and Zr^{4+}, whereas Ru was identified as Ru^0 except for the spectra near the surface region (depth < 13 nm), where the Ru^{x+} and Ru^{4+} signals were split. The binding energy value of Ru^0 $3d_{5/2}$ (279.96 ± 0.08 eV) was consistent with that of other coatings (279.69–280.16 eV) reported in the literature [13,16,17,27], whereas the binding energies of Ru^{x+} and Ru^{4+} $3d_{5/2}$ were 280.45 ± 0.11 and 282.57 ± 0.15 eV, respectively. Previous studies reported 281.4–282.2 eV [26,28–30] for the binding energy of Ru^{4+} $3d_{5/2}$. Ru of 17%–20% exhibited the Ru^{4+} state at a depth of 0–13 nm. Ru atoms remained in its metallic state beneath the near surface region. Figure 6d shows the intensity variations of O^{2-} 1s, Ru^0 $3d_{5/2}$, and Zr^{4+} $3d_{5/2}$ signals along the depth, which indicates that the variation trend of the

O^{2-} 1s profile coincides with that of the Zr^{4+} $3d_{5/2}$ profile and is in contrast to that of the Ru^0 $3d_{5/2}$ profile, implying that ZrO_2 is the dominant oxide. Therefore, the annealed $Ru_{0.50}Zr_{0.50}$(R1) coating comprised alternating oxygen-rich and oxygen-deficient layers stacked along the O-diffusion direction. The binding energy value of Zr^{4+} $3d_{5/2}$ deviated within 182.05–183.35 eV (Figure 6e). Moreover, this range decreased to 182.71–183.35 eV after the data in the first laminated period had been excluded. Previous studies have reported 182.75 [31], 182.8 [32], and 182.9 eV [33] for the binding energy value of Zr^{4+} $3d_{5/2}$. The binding energy value of O^{2-} 1s demonstrated a variation pattern similar to that of the binding energy value of Zr^{4+} $3d_{5/2}$ (Figure 6e). The charging effect of analyzing insulators [34] caused substantial deviation in binding energy. The binding energy difference $\Delta = (O^{2-}\ 1s - Zr^{4+}\ 3d_{5/2})$ was 347.92 ± 0.05 eV at the analyzed depth of 19.5–318.5 nm. This difference was highly consistent with the reported difference of 348.01 and 348.2 eV, calculated using 530.76 and 182.75 eV [31] or 531.1 and 182.9 eV [33] for O^{2-} 1s and Zr^{4+} $3d_{5/2}$, respectively. The periodic changes of nonoxidized metallic Ru suggested the influence of oxygen in the Zr-deficient sublayers.

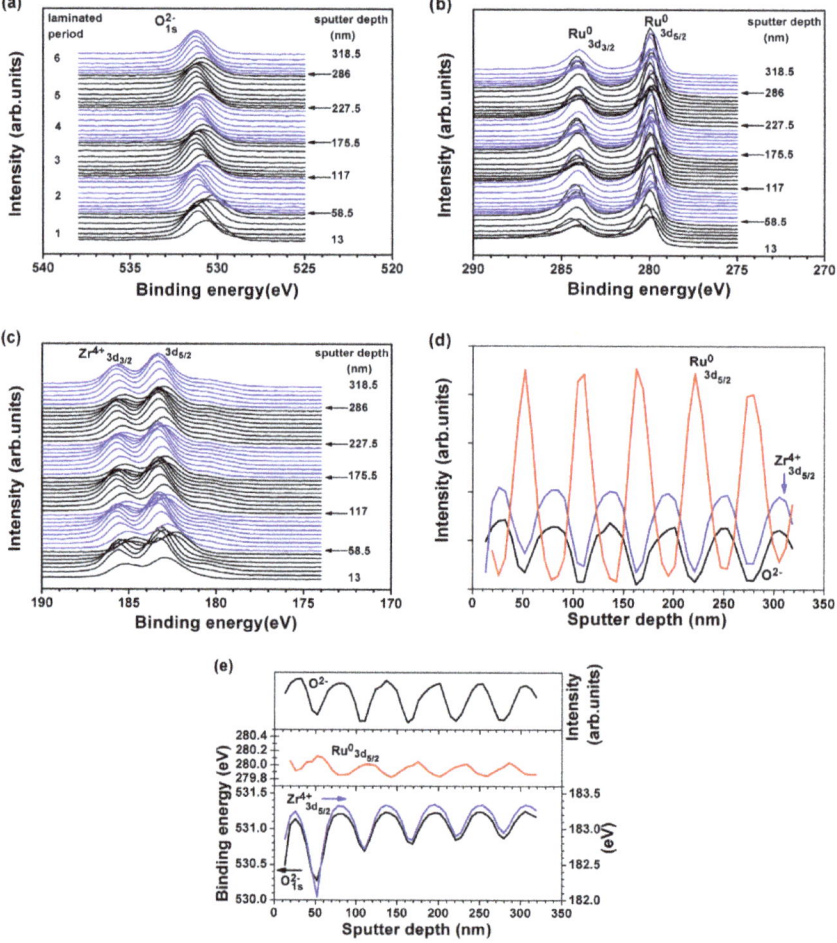

Figure 6. XPS spectra of (**a**) O 1s, (**b**) Ru 3d, and (**c**) Zr 3d core levels of the $Ru_{0.50}Zr_{0.50}$(R1) coating after annealing in 1% O_2–99% Ar at 600 °C for 30 min; variation patterns of (**d**) intensity and (**e**) binding energy of O^{2-} 1s, $Zr^{4+}3d_{5/2}$, and Ru^0 $3d_{5/2}$.

Figure 7a,b shows the cross-sectional TEM images of the annealed $Ru_{0.48}Zr_{0.52}$(R5) coating; the laminated structure was evident. Figure 7c shows a high-resolution TEM image of the near-surface region of the annealed coating. The lattice fringes of particular areas indicated that the annealed $Ru_{0.48}Zr_{0.52}$(R5) coating comprised ZrO_2- and Ru-dominant sublayers, which linked together across the original columnar boundaries such that the annealed $Ru_{0.48}Zr_{0.52}$(R5) coatings were laminated and the columnar boundaries were unresolved. Figure 8a depicts the cross-sectional TEM image of the annealed $Ru_{0.47}Zr_{0.53}$(R10) coating. The laminated sublayers were curved, because of which the stacks of sublayers among neighboring columnar structures were disconnected. Figure 8b shows the high-resolution TEM image of the middle region of the annealed $Ru_{0.47}Zr_{0.53}$(R10) coating. The Ru-dominant sublayers were two-nanometers thick only, and disconnected regions of the sublayers among neighboring columnar structures were observed. The fast variation of cyclical gradient concentration for the R10 coatings prepared with a quick substrate holder rotation speed resulted in the formation of grooved sublayers. For the coatings with thicker Ru sublayers, R1, R3, and R5, the sublayers became flat. The misaligned connections were more evident in the near-surface region (Figure 8c).

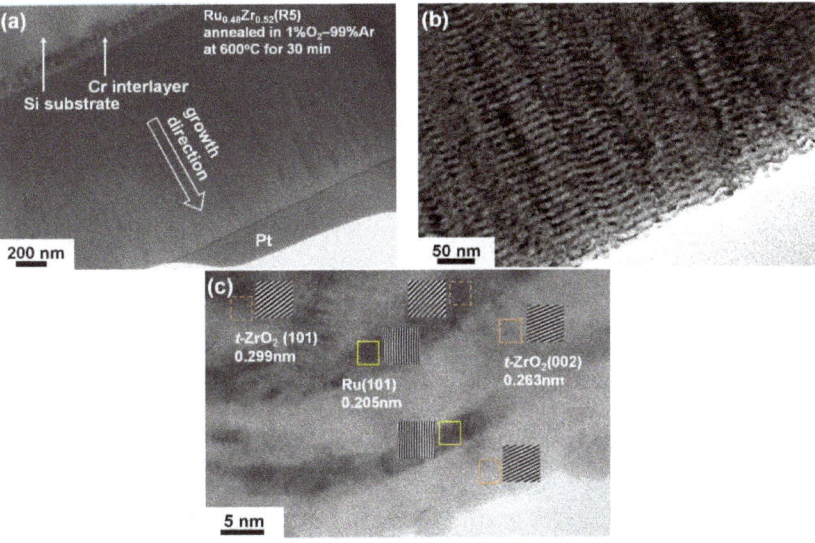

Figure 7. (a,b) cross-sectional TEM images of the oxidized $Ru_{0.48}Zr_{0.52}$(R5) coating; (c) high-resolution image of the near surface region of the annealed $Ru_{0.48}Zr_{0.52}$(R5) coating.

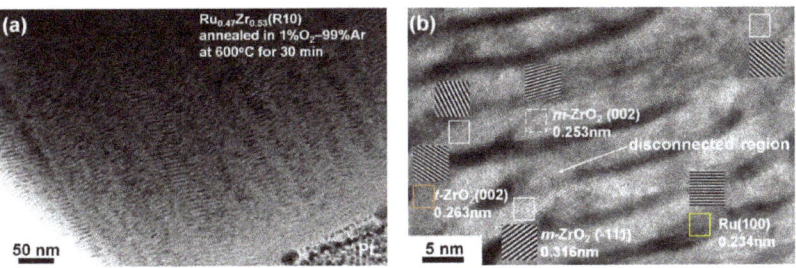

Figure 8. Cont.

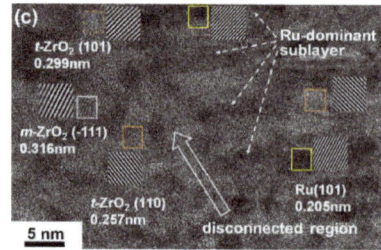

Figure 8. (a) Cross-sectional TEM image of the oxidized $Ru_{0.47}Zr_{0.53}$(R10) coating; high-resolution images of the (b) middle region and (c) near-surface region of the annealed coating.

Figure 9a shows a cross-sectional TEM image of the annealed $Ru_{0.46}Zr_{0.54}$(R30) coating, in which the original columnar boundaries are evident, but no laminated structures were observed. A high-resolution TEM image (Figure 9b) revealed nanocrystalline grains of ZrO_2 and Ru, each approximately five nanometers in diameter, implying that a nanocomposite structure had been constructed. Furthermore, Ru grains, the dark regions in the image, tended to concentrate along the columnar boundaries. Figure 10a–c illustrates the XPS spectra of the annealed $Ru_{0.46}Zr_{0.54}$(R30) coating. The XPS spectra of Ru 3d core levels indicated the presence of minor Ru^{4+} ($3d_{5/2}$: 282.11 eV) in addition to Ru^0 ($3d_{5/2}$: 280.19 ± 0.07 eV) at the near-surface region (Figure 10b), which was attributed to the incorporation of Ru into the ZrO_2 grains because RuO_2 and ZrO_2 possessed a similar tetragonal structure. Figure 10d shows that the intensities of O^{2-} 1s, Ru^0 $3d_{5/2}$, and Zr^{4+} $3d_{5/2}$ signals were constant along the depth due to the limit of XPS analyses. Similar binding energy trends were observed for O^{2-} and Zr^{4+} (Figure 10e). The binding energy difference $\Delta = (O^{2-}$ 1s $- Zr^{4+}$ $3d_{5/2})$ was 348.00 ± 0.02 eV at the analyzed depth (5.7–96.9 nm). Therefore, Zr reacted with O during annealing, and the annealed coating exhibited a nanocomposite comprising ZrO_2 and Ru grains.

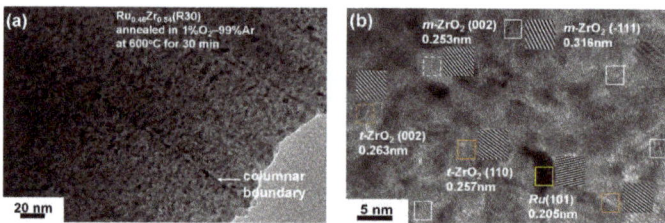

Figure 9. (a) Cross-sectional TEM image and (b) high-resolution image of the oxidized $Ru_{0.46}Zr_{0.54}$(R30) coating.

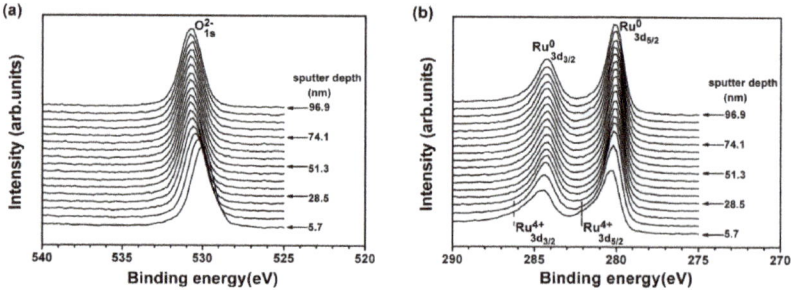

Figure 10. *Cont.*

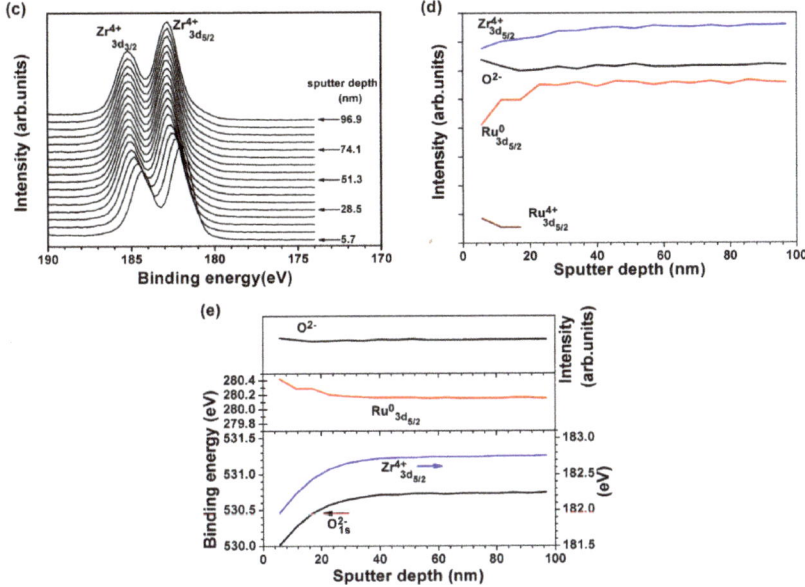

Figure 10. XPS spectra of (**a**) O 1s, (**b**) Ru 3d, and (**c**) Zr 3d core levels of the $Ru_{0.46}Zr_{0.54}$(R30) coating after annealing in 1% O_2–99% Ar at 600 °C for 30 min; variation patterns of (**d**) intensity and (**e**) binding energy of O^{2-} 1s, $Zr^{4+}3d_{5/2}$, and $Ru^0\ 3d_{5/2}$.

3.3. Mechanical Properties of Internally Oxidized Ru–Zr Coatings

Figure 11 depicts the nanoindentation hardness variations of the as-deposited and internally oxidized Ru–Zr coatings prepared at various substrate holder rotation speeds through sputtering. The hardness of the as-deposited coatings increased from 9.1 to 13.1 GPa with the substrate holder rotation speed and decreasing equilibrated laminated layer period. This hardness increase was attributed to the decrease in crystalline size and structural variation. The nanoindentation hardness of all Ru–Zr coatings increased after annealing in 1% O_2–99% Ar at 600 °C for 30 min. The hardness variation curve of the internally oxidized Ru–Zr coatings exhibited three divisions representing a laminated structure, a disconnected laminated structure, and a nanocomposite region. The hardness increased from 9.1, 10.3, and 10.5 to 15.5, 16.1, and 17.2 GPa for the annealed $Ru_{0.50}Zr_{0.50}$(R1), $Ru_{0.49}Zr_{0.51}$(R3), and $Ru_{0.48}Zr_{0.52}$(R5) coatings, respectively, which exhibited equilibrated laminated layer periods of 55, 18, and 11 nm, respectively. This result indicates that the hardness of the internally oxidized Ru–Zr coatings, which exhibited crystalline phases identical to those identified through XRD analysis and appropriately maintained their multilayer structures, was affected by the layer period. These internally oxidized Ru–Zr multilayer coatings were categorized as nonisostructural oxide/metal multilayers [1]. Dislocation could not be moved across oxide/metal interfaces because oxides are brittle materials that deform through fracture mechanisms, limiting the hardness enhancement [2]; therefore, the hardness of oxide/metal multilayers approached that of the oxide ZrO_2. Gan et al. reported a nanoindentation hardness of 18 GPa for monoclinic ZrO_2 thin films [35]. By contrast, the hardness of the annealed $Ru_{0.47}Zr_{0.53}$(R10) coatings with an equilibrated laminated layer period of 5.6 nm exhibited a relatively low level of 12.3 GPa. Although the internally oxidized $Ru_{0.47}Zr_{0.53}$(R10) coatings were laminated in each columnar structure, the same sublayers among neighboring columnar structures were misaligned and disconnected, which reduced the hardness. The internally oxidized $Ru_{0.46}Zr_{0.54}$(R15), $Ru_{0.47}Zr_{0.53}$(R20), and $Ru_{0.46}Zr_{0.54}$(R30) coatings exhibited high hardness within 16.1–17.9 GPa and were categorized as nanocrystalline composites consisting of hard ZrO_2 grains and soft Ru grains. Figure 12 shows the variation in Young's moduli of the as-deposited and internally oxidized Ru–Zr coatings. The Young's moduli increased from 130 to 140 GPa for R1, R3, and R5 coatings, to 160 GPa for R10 coatings and 170–180 GPa for R15, R20, and R30 coatings. Because the internally oxidized Ru–Zr coatings exhibited similar phases, ZrO_2 and Ru, the differences in Young's moduli among the annealed coatings were limited (i.e., 160–180 GPa). The surface roughness values of the Ru–Zr coatings are shown in Table 1. When evaluating the mechanical properties of coatings, previous studies [36–38] have reported that coatings with higher surface roughness exhibit larger standard deviation values or lower mean values. The effect of surface roughness on the mechanical properties of as-deposited Ru–Zr coatings was unclear. By contrast, the mechanical properties of the annealed coatings revealed larger deviations and higher surface roughness values than did those of the as-deposited coatings.

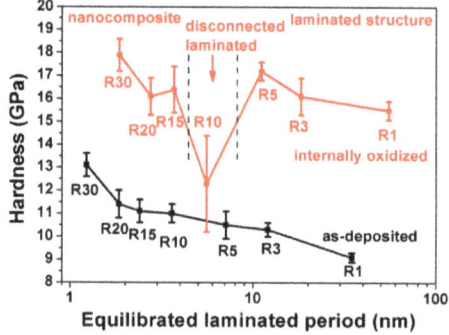

Figure 11. Nanoindentation hardness values of the as-deposited and internally oxidized Ru–Zr coatings.

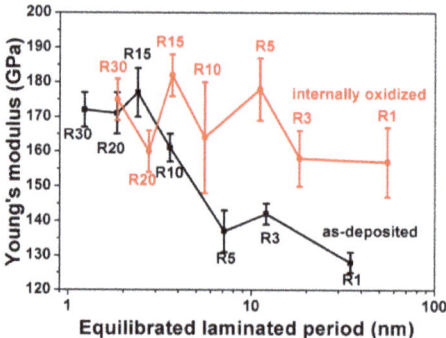

Figure 12. Young's modulus values of the as-deposited and internally oxidized Ru–Zr coatings.

4. Conclusions

Rotation speeds of the substrate holder during sputtering affected the crystalline structure and mechanical properties of Ru–Zr coatings both in the as-deposited and internally oxidized states. Because Ru–Zr coatings were fabricated using a cyclical gradient concentration stacked constitution, the coatings prepared at low rotation speeds (1–10 rpm) exhibited a laminated structure in addition to a columnar structure. The as-deposited Ru–Zr coatings exhibited nanoindentation hardness of 9.1–13.1 GPa, and the coatings prepared at higher substrate holder rotation speeds exhibited higher hardness. After annealing in a 1% O_2–99% Ar atmosphere at 600 °C for 30 min accompanied by the conduction of internal oxidation, the coatings prepared at a substrate holder rotation speed of one to five revolutions per minute maintained a laminated structure; this structure comprised alternately stacked Ru-dominant and ZrO$_2$-dominant sublayers whose nanoindentation hardness increased to 15.5–17.2 GPa because of the formation of ZrO$_2$ phase and the maintenance of sublayer interfaces. By contrast, the annealed coatings prepared at a rotation speed of 10 rpm maintained a similar laminated structure; however, the stacks of sublayers among neighboring columnar structures were misaligned and disconnected, resulting in relatively low nanoindentation hardness of 12.3 GPa. The annealed coatings prepared at a substrate holder rotation speed of 15–30 rpm exhibited nanocomposite coatings comprising Ru and ZrO$_2$ grains within evident columnar boundaries and a high nanoindentation hardness of 16.1–17.9 GPa.

Acknowledgments: The financial support of this work from the Ministry of Science and Technology, Taiwan, under Contract No. 102-2221-E-019-007-MY3 is appreciated.

Author Contributions: Yung-I Chen designed the experiments and wrote the paper; Tso-Shen Lu performed the experiments; Zhi-Ting Zheng analyzed the XPS data.

Conflicts of Interest: The authors declare no conflict of interest. The founding sponsors had no role in the design of the study; in the collection, analyses, or interpretation of data; in the writing of the manuscript, and in the decision to publish the results.

References

1. Yashar, P.C.; Sproul, W.D. Nanometer scale multilayered hard coatings. *Vacuum* **1999**, *55*, 179–190. [CrossRef]
2. Yashar, P.C.; Barnett, S.A.; Hultman, L.; Sproul, W.D. Deposition and mechanical properties of polycrystalline Y$_2$O$_3$/ZrO$_2$ superlattices. *J. Mater. Res.* **1999**, *14*, 3614–3622. [CrossRef]
3. Vüllers, F.T.N.; Spolenak, R. From solid solutions to fully phase separated interpenetrating networks in sputter deposited "immiscible" W–Cu thin films. *Acta Mater.* **2015**, *99*, 213–227. [CrossRef]
4. Beainou, R.E.; Martin, N.; Potin, V.; Pedrosa, P.; Yazdi, M.A.P.; Billard, A. Correlation between structure and electrical resistivity of W–Cu thin films prepared by GLAD co-sputtering. *Surf. Coat. Technol.* **2017**, *313*, 1–7. [CrossRef]
5. Müller, C.M.; Sologubenko, A.S.; Gerstl, S.S.A.; Süess, M.J.; Courty, D.; Spolenak, R. Nanoscale Cu/Ta multilayer deposition by co-sputtering on a rotating substrate. Empirical model and experiment. *Surf. Coat. Technol.* **2016**, *302*, 284–292. [CrossRef]
6. Müller, C.M.; Spolenak, R. An in situ X-ray diffraction study of phase separation in Cu–Ta alloy thin films. *Thin Solid Films* **2016**, *598*, 276–288. [CrossRef]
7. Zotov, N.; Woll, K.; Mücklich, F. Phase formation of B2-RuAl during annealing of Ru/Al multilayers. *Intermetallics* **2010**, *18*, 1507–1516. [CrossRef]
8. Guitar, M.A.; Aboulfadl, H.; Pauly, C.; Leibenguth, P.; Migot, S.; Mücklich, F. Production of single-phase intermetallic films from Ru-Al multilayers. *Surf. Coat. Technol.* **2014**, *244*, 210–216. [CrossRef]
9. Manhardt, H.; Lupton, D.F.; Kock, W. Gold-Free Platinum Material Dispersion-Strengthened by Small, Finely Dispersed Particles of Base Metal Oxide. U.S. Patent 6,663,728, 16 December 2003.
10. Nakamura, T.; Sakaguchi, O.; Kusamori, H.; Matsuzawa, O.; Takahashi, M.; Yamamoto, T. Method for Preparing Ag-ZnO Electric Contact Material and Electric Contact Material Produced Thereby. U.S. Patent 6,432,157, 13 August 2002.
11. Khanna, A.S. *Introduction to High Temperature Oxidation and Corrosion*; ASM International: Materials Park, OH, USA, 2002.
12. Chen, Y.I.; Chang, L.C.; Huang, R.T.; Tsai, B.N.; Kuo, Y.C. Internal Oxidation of Mo–Ru Coatings. *Thin Solid Films* **2010**, *518*, 3819–3824. [CrossRef]
13. Chen, Y.I.; Tsai, B.N. Annealing and oxidation study of Ta–Ru hard coatings. *Surf. Coat. Technol.* **2010**, *205*, 1362–1367. [CrossRef]
14. Chen, Y.I. Laminated structure in internally oxidized Ru–Ta coatings. *Thin Solid Films* **2012**, *524*, 205–210. [CrossRef]
15. Chen, Y.I.; Chu, H.N.; Chang, L.C.; Lee, J.W. Internal oxidation and mechanical properties of Ru based alloy coatings. *J. Vac. Sci. Technol. A* **2014**, *32*, 02B101. [CrossRef]

16. Chen, Y.I.; Lu, T.S. Internal oxidation of laminated ternary Ru–Ta–Zr coatings. *Appl. Surf. Sci.* **2015**, *353*, 245–253. [CrossRef]
17. Chen, Y.I.; Chu, H.N.; Kai, W. Internal oxidation of laminated Nb–Ru coatings. *Appl. Surf. Sci.* **2016**, *389*, 477–483. [CrossRef]
18. Saha, R.; Nix, W.D. Effects of the substrate on the determination of thin film mechanical properties by nanoindentation. *Acta Mater.* **2002**, *50*, 23–38. [CrossRef]
19. Oliver, W.C.; Pharr, G.M. An improved technique for determining hardness and elastic modulus using load and displacement sensing indentation experiments. *J. Mater. Res.* **1992**, *7*, 1564–1583. [CrossRef]
20. Bennett, J.M. *Rough Surfaces*, 2nd ed.; Imperial College Press: London, UK, 1999.
21. Mahdouk, K.; Elaissaoui, K.; Charles, J.; Bouirden, L.; Gachon, J.C. Calorimetric study and optimization of the ruthenium-zirconium phase diagram. *Intermetallics* **1997**, *5*, 111–116. [CrossRef]
22. Wang, F.E. Equiatomic binary compounds of Zr with transition elements Ru, Rh, and Pd. *J. Appl. Phys.* **1967**, *38*, 822–824. [CrossRef]
23. David, N.; Benlaharche, T.; Fiorani, J.M.; Vilasi, M. Thermodynamic modeling of Ru–Zr and Hf–Ru systems. *Intermetallics* **2007**, *15*, 1632–1637. [CrossRef]
24. Arikan, N.; Bayhan, Ü. *Ab initio* calculation of structural, electronic and phonon properties of ZrRu and ZrZn in B2 phase. *Phys. B* **2011**, *406*, 3234–3237. [CrossRef]
25. Xing, W.; Chen, X.Q.; Li, D.; Li, Y.; Fu, C.L.; Meschel, S.V.; Ding, X. First-principles studies of structural stabilities and enthalpies of formation of refractory intermetallics: TM and TM3 (T = Ti, Zr, Hf; M = Ru, Rh, Pd, Os, Ir, Pt). *Intermetallics* **2012**, *28*, 16–24. [CrossRef]
26. Chen, Y.I.; Zheng, Z.T.; Kai, W.; Huang, Y.R. Oxidation behavior of Ru–Al multilayer coatings. *Appl. Surf. Sci.* **2017**, *406*, 1–7. [CrossRef]
27. Chen, Y.I.; Chang, L.C.; Lee, J.W.; Lin, C.H. Annealing and oxidation study of Mo–Ru hard coatings on tungsten carbide. *Thin Solid Films* **2009**, *518*, 194–200. [CrossRef]
28. Rochefort, D.; Dabo, P.; Guay, D.; Sherwood, P.M.A. XPS investigations of thermally prepared RuO_2 electrodes in reductive conditions. *Electrochim. Acta* **2003**, *48*, 4245–4252. [CrossRef]
29. Chen, Y.I.; Chen, S.M.; Chang, L.C.; Chu, H.N. X-ray photoelectron spectroscopy and transmission electron microscopy study of internally oxidized Nb–Ru coatings. *Thin Solid Films* **2013**, *544*, 491–495. [CrossRef]
30. Cox, P.A.; Goodenough, J.B.; Tavener, P.J.; Telles, D.; Egdell, R.G. The electronic structure of $Bi_{2-x}Gd_xRu_2O_7$ and RuO_2: A study by electron spectroscopy. *J. Solid State Chem.* **1986**, *62*, 360–370. [CrossRef]
31. Morant, C.; Sanz, J.M.; Galán, L.; Soriano, L.; Rueda, F. An XPS study of the interaction of oxygen with zirconium. *Surf. Sci.* **1989**, *218*, 331–345. [CrossRef]
32. Maurice, V.; Salmeron, M.; Somorjai, G.A. The epitaxial growth of zirconium oxide thin films on Pt (111) single crystal surfaces. *Surf. Sci.* **1990**, *237*, 116–126. [CrossRef]
33. Wang, Y.M.; Li, Y.S.; Wong, P.C.; Mitchell, K.A.R. XPS studies of the stability and reactivity of thin films of oxidized zirconium. *Appl. Surf. Sci.* **1993**, *72*, 237–244. [CrossRef]
34. Ramana, C.V.; Atuchin, V.V.; Kesler, V.G.; Kochubey, V.A.; Pokrovsky, L.D.; Shutthanandan, V.; Becker, U.; Ewing, R.C. Growth and surface characterization of sputter-deposited molybdenum oxide thin films. *Appl. Surf. Sci.* **2007**, *253*, 5368–5374. [CrossRef]
35. Gan, Z.; Yu, G.; Zhao, Z.; Tan, C.M.; Tay, B.K. Mechanical properties of zirconia thin films deposited by filtered cathodic vacuum arc. *J. Am. Ceram. Soc.* **2005**, *88*, 2227–2229. [CrossRef]
36. Qasmi, M.; Delobelle, P. Influence of the average roughness R_{ms} on the precision of the Young's modulus and hardness determination using nanoindentation technique with a Berkovich indenter. *Surf. Coat. Technol.* **2007**, *201*, 1191–1199. [CrossRef]
37. Walter, C.; Antretter, T.; Daniel, R.; Mitterer, C. Finite element simulation of the effect of substrate roughness on nanoindentation of thin films with spherical indenters. *Surf. Coat. Technol.* **2007**, *202*, 1103–1107. [CrossRef]
38. Kim, J.Y.; Kang, S.K.; Lee, J.J.; Jang, J.; Lee, Y.H.; Kwon, D. Influence of surface-roughness on indentation size effect. *Acta Mater.* **2007**, *55*, 3555–3562. [CrossRef]

© 2017 by the authors. Licensee MDPI, Basel, Switzerland. This article is an open access article distributed under the terms and conditions of the Creative Commons Attribution (CC BY) license (http://creativecommons.org/licenses/by/4.0/).

Article

Corrosion Protection of Steel by Epoxy-Organoclay Nanocomposite Coatings

Domna Merachtsaki, Panagiotis Xidas, Panagiotis Giannakoudakis,
Konstantinos Triantafyllidis * and Panagiotis Spathis *

Department of Chemistry, Aristotle University of Thessaloniki, University Campus, Thessaloniki 54124, Greece; do7mera@gmail.com (D.M.); pxidas@chem.auth.gr (P.X.); panjian@chem.auth.gr (P.G.)
* Correspondence: ktrianta@chem.auth.gr (K.T.); spathis@chem.auth.gr (P.S.);
 Tel.: +3-23-1099-7730 (K.T.); +3-23-1099-7835 (P.S.)

Received: 16 March 2017; Accepted: 13 June 2017; Published: 22 June 2017

Abstract: The purpose of the present work was to study the corrosion behavior of steel coated with epoxy-(organo) clay nanocomposite films. The investigation was carried out using salt spray exposures, optical and scanning electron microscopy examination, open circuit potential, and electrochemical impedance measurements. The mechanical, thermomechanical, and barrier properties of pristine glassy epoxy polymer and epoxy-clay nanocomposites were examined. The degree of intercalation/exfoliation of clay nanoplatelets within the epoxy polymer also was determined. The mechanical, thermomechanical, and barrier properties of all the epoxy-clay nanocomposites were improved compared to those of the pristine epoxy polymer. In addition, both the pristine epoxy and the epoxy nanocomposite coatings protected the steel from corrosion. Furthermore, the protective properties of the nanocomposite coatings were superior compared to those of the pristine epoxy polymer. The protective properties of the nanocomposite coatings varied with the modified clay used. The epoxy-montmorillonite clay modified with primary octadecylammonium ions, Nanomer I.30E, had a better behavior than that modified with quaternary octadecylammonium ions, Nanomer I.28E.

Keywords: steel; corrosion; protection; coatings; epoxy—clay nanocomposites

1. Introduction

Organic coatings are widely used to prevent corrosion of metallic structures. However, these polymeric coatings are usually permeable to small gaseous molecules such as water vapor and oxygen, which can result in gradual corrosion of the surface. It is generally accepted that the coating efficiency is dependent on the barrier and mechanical (resistance to cracking) properties of the organic film, on the adherence of the polymeric coating to the metal substrate, and on the degree of environmental aggressiveness. Among various protective coatings, epoxy resins are commonly used as organic coatings, due to their strong adhesion capability to metallic substrates, their excellent resistance to chemicals, and their relatively high mechanical and impact strength. However, the pristine epoxy resins exhibit measureable adsorption and permeability of water vapor, which diffuses to the epoxy/steel interface and initiates corrosion of the metal substrate particularly in intensely wet conditions. Therefore, effort has been devoted in recent years to develop epoxy-based protective coatings with good barrier properties, at least with regard to water vapor and oxygen [1–3].

One way to improve the properties of a polymeric protective coating is the addition of inorganic nano-fillers, leading to the formation of nanocomposite materials. Nanoparticles with sizes in the range of ca. 1–50 nm can enhance the effectiveness of a coating by filling the micro voids and crevices in the pristine polymer coating. When the nanoparticles are in the form of 2D-nanolayers with a high aspect ratio (ca. 100–2000), they can act as barriers to the diffusion of small molecules by increasing the length of the diffusion paths (tortuous paths) of the corroding agent through the polymeric nanocomposite coating, thus inhibiting the corrosion process. The use of clays as fillers of polymeric coatings has great potential to improve the barrier properties of the coatings, provided that the nano-sized clay tactoids or the individual clay nanolayers can be dispersed effectively within the polymer matrix. However, substantial improvements of barrier properties can be achieved when the nanolayers are oriented parallel to the substrate surface [4].

During the last 20 years, considerable attention has been paid to the development of polymer-based nanocomposites, with clays being the first and most studied inorganic nanofillers used [5,6]. It has been reported that the incorporation of a small amount (1–5 wt %) of clay in a polymer matrix can lead to significant improvements in the mechanical performance, thermal stability, and barrier properties of the pristine polymer. These improvements are related to the morphology of clay micro-sized particles, which consists of tactoids with highly oriented nano-layers, as explained above. An ideal exfoliated structure of polymer-clay nanocomposites is obtained when a complete separation and dispersion of the individual clay nano-layers occurs within the polymer matrix. In this case, there is no longer any interaction between the layers, which are completely dissociated and separated by a large volume of the polymer. An intermediate case is the intercalated structure where a finite number of polymer chains penetrate the interlayer space, thus significantly increasing the spacing between the layers (ca. up to 100 Å) without destroying the ordered parallel structure of the nanolayers, at least at the level of individual tactoids.

The most widely used layered silicate is montmorillonite (MMT), which has attracted intense research interest for the preparation of polymer clay nanocomposites. The MMT-based nanocomposites exhibit enhanced physical properties compared to the pristine polymer, such as improved thermal properties (e.g., thermal stability, flame retardant, thermal conductivity), mechanical properties (e.g., mechanical strength, hardness, abrasion resistance), permeability properties (e.g., gas barrier, pervaporation), and corrosion protection properties [1,2,7–11]. The chemical structure of MMT consists of two tetrahedral silica sheets fused to a central edge-shared octahedral-based sheet of either magnesium or aluminum hydroxide [12]. In general, the surface of the clay needs to be organo-modified in order to become more compatible with the polymer matrix and to improve its dispersibility in the polymeric network. The organic modification of layered silicates can be realized through the replacement of the Na^+ and/or Ca^{2+} cations in the intragallery space, as well as on the external surfaces of the clay particles, by organic cations through a cation exchange reaction [13,14]. The improvement in the corrosion resistance of aluminium alloys and of cold rolled steel with polymeric films reinforced with organically modified clay has been clearly demonstrated. Corrosion protection was essentially related to the enhancement of the barrier properties of the coating. The formation of an organophilic environment between the clay layers is critical for the insertion of polymer chains amongst them and the formation of an intercalated structure, which seems to present the greater improvement of barrier properties. Amongst the various polymers for coating applications, epoxy resins lately have evoked intensive studies in the preparation of nanocomposite materials, due to their high tensile strength and modulus, good adhesive properties, good chemical and corrosion resistance, low shrinkage in cure, and excellent dimensional stability [1,8,15–19].

In the present study, the protection capabilities of epoxy-clay nanocomposite coatings were examined. The montmorillonite clay used has been modified with quaternary or primary octadecylammonium ions. Both the pristine glassy epoxy polymer and the epoxy-clay nanocomposites were characterized for their mechanical and thermomechanical properties, thermal stability, and barrier properties. The degree of intercalation/exfoliation of clay nanoplatelets within the epoxy polymer was determined. The corrosion behavior investigation was carried out using salt spray tests, optical and scanning electron microscopy examination, open circuit potential, and electrochemical impedance measurements.

2. Materials and Methods

2.1. Materials and Epoxy Coatings

The steel test material was cold rolled steel (DC 01—ASTM A366 [20] with a chemical composition max %, 0.12 C, 0.045 P, 0.045 S, 0.60 Mn, and the specimens were cut from a plate of 0.3 cm thickness. The dimensions of the test coupons were 5 cm × 1.5 cm, the total exposed area for the salt spray tests was 15 cm^2, and for the electrochemical measurements was 6 cm^2. The steel surface before coating was mechanically cleaned by scrubbing with a bristle brush and chemically cleaned using acetone and alcohol.

The metallic specimens were coated with a ~20 µm thin film of pristine glassy epoxy polymer or epoxy-clay nanocomposite films. No pinholes or other defects were observed on the coatings. The pristine liquid epoxy resin was diglycidyl ether of bisphenol A (DGEBA) (EPON 828RS, Hexion, Columbus, OH, USA) with an average epoxide equivalent weight of ~187 (M_W = 370) and was mixed at 50 °C with the appropriate amount of an aliphatic polyoxypropylene diamine (Jeffamine D-230, $M_W \approx 230$, Huntsman, The Woodlands, TX, USA), which acted as the curing agent. The molecular structures of the epoxy resin and diamine curing agent are shown in Figure 1A,B, respectively.

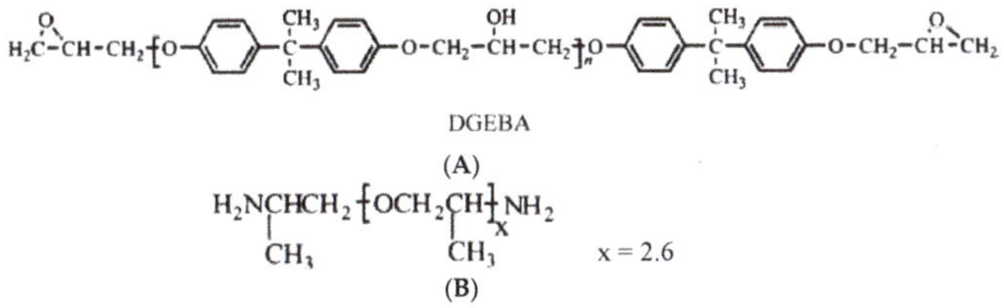

Figure 1. Molecular structure of (**A**) diglycidyl ether of bisphenol-A (DGEBA) and (**B**) of aliphatic polyoxypropylene diamine (Jeffamine®).

The curing agent was mixed with the epoxy resin under a stoichiometric ratio of 1:1 of amine reactive hydrogens to epoxy rings. The metallic specimens were dipped in the liquid uncured mixture and then were kept in a vertical position so that the excess liquid was removed and left to cure at ambient conditions for 24 h. Final post-curing was performed at 75 °C for 3 h and 125 °C for another 3 h. The same procedure was applied for the coating of the specimens with the epoxy-clay nanocomposites, except that prior to adding the curing agent, the epoxy pre-polymer (DGEBA) was mixed with the (organo) clay for 1 h at 50 °C. The clay loading in the nanocomposites was 3 and 6 wt % on a silicate basis.

The clays used for preparing the nanocomposite coating were the Nanomer I.28E and the Nanomer I.30E, both kindly provided by Nanocor Inc. (Hoffman Estates, IL, USA), which are montmorillonite clays that have been modified with quaternary and primary octadecylammonium ions, respectively. The parent, inorganic Na$^+$-PGW clay (Polymer Grade Wyoming, Nanocor Inc.) was also used for preparing specimens of epoxy-clay nanocomposites for comparing the structure and properties of the bulk nanocomposites samples. All three clays used were polymer grade (PG) montmorillonites which are high purity aluminosilicate minerals with the theoretical chemical formula: $M^+{}_y(Al_{2-y} Mg_y)(Si_4)O_{10}(OH)_2 \cdot nH_2O$. The chemical composition, as measured by Inductively Coupled Plasma Atomic Emission Spectroscopy, ICP-AES, chemical analysis, of the Na$^+$-PGW montmorillonite clay and the two organoclays is presented in Table 1 below. Obviously, there is a dramatic reduction of Na$^+$ cations in the composition of the organoclays.

The average particle size of Na-PGW as measured by a laser particle size analyzer (Mastersizer S, Malvern Instruments, Malvern, UK) in 0.4 wt % aqueous suspensions was ~2 µm (with a distribution of 0.5–10 µm). Various relevant physicochemical properties of the clays used, as provided by Nanocor Inc., are given in Table 2.

Table 1. Data of ICP-AES chemical analysis of parent sodium montmorillonite PGW, I.28E quarternary octadecyl ammonium organoclay, and I.30E primary octadecyl ammonium organoclay.

Samples Were Calcined at 600 °C	Aluminum % (Al)	Calcium % (Ca)	Iron % (Fe)	Potassium % (K)	Magnesium % (Mg)	Sodium % (Na)
Na$^+$-PGW	12.02	0.36	1.57	0.16	2.3	3.31
I.30E	12.48	0.28	1.7	0.17	2.37	0.32
I.28E	12.17	0.22	1.74	0.23	2.3	0.08

Table 2. Physical properties of Nanocor Inc. montmorillonite clays.

Property	Na$^+$-PGW	I.30E	I.28E
Color	White Powder	White Powder	White Powder
Cation Exchange Capacity, CEC (meq/100 g) ± 10%	145	–	–
Mean Dry Particle Size (μm)	~2 (0.5–10)	8–10	8–10
Aspect Ratio	200–400	–	–
+325 Mesh Residue (%)	–	0.1	0.1
Specific Gravity	2.6	1.71	1.9
Max Moisture (%)	12	3	3
pH (5% dispersion)	9.5–10.5	–	–
Bulk Density (pounds/ft3) (gm/cc)	–	250.41	260.42
Mineral Purity (min %)	–	98.5	98.5

Notes: Nanocor Inc. Vol. Lit. G-105 "POLYMER GRADE MONTMORILLONITES" (Nanocor, 2006); Nanocor Inc. Vol. Lit. T-11 "Epoxy Nanocomposites Using Nanomer® I.30E Nanoclay" (Nanocor, 2004); Nanocor Inc. Vol. Lit. T-12 "Nanocomposites Using Nanomer® I.28E Nanoclay" (Nanocor, 2004).

2.2. Characterization and Testing of Epoxy Polymer and Nanocomposites

Both the pristine glassy epoxy polymer and the epoxy-clay nanocomposites were characterized for their mechanical properties (tensile measurements), thermomechanical properties (Dynamic Mechanical Analysis, DMA), thermal stability (Thermogravimetric Analysis, TGA), and barrier properties (O_2 permeability measurements). X-ray diffraction (XRD) was also applied in order to determine the degree of intercalation/exfoliation of clay nanoplatelets within the epoxy polymer. The dispersion of clay nanolayers within the polymer matrix was investigated by means of High Resolution Transmission Electron Microscopy (HRTEM).

XRD measurements were performed on a Siemens D-500 (Siemens AG, Karlsruhe, Germany) type automated diffractometer, with Cu(Kα) λ = 1.5418 Å radiation, in the range 2°–10° 2θ and at a scan rate of 1°/min. HRTEM measurements were performed on a JEOL 2011 high resolution transmission electron microscope (Jeol, Peabody, MA, USA) with a LaB$_6$ filament (TED PELLA Inc., Redding, CA, USA) and an accelerating voltage of 200 kV, a point resolution of 0.23 nm, and a spherical aberration coefficient of C_s = 1 mm. The TEM samples were prepared by supporting thin sections (80–100 nm) of the nanocomposite samples onto a lacy carbon film supported on a 3 mm diameter, 300 mesh copper grid. The specimens were further coated with a carbon layer in order to enhance conductivity and avoid destruction of the epoxy polymer.

The mechanical properties of the samples were measured with an Instron 3344 dynamometer (Instron, Norwood, MA, USA) according to the Standard Method (ASTM D638) [21], with a stress rate of 5 mm/min. The specimens had a dogbone shape and dimensions of 40 mm × 5 mm × 2 mm and were prepared in rubber molds by curing of the epoxy/amine mixture, as described above. The thermomechanical properties were measured using a Perkin Elmer Diamond DMA analyser (Perkin Elmer, Waltham, MA, USA). The bending method was used with a frequency of 1 Hz in a temperature range 25–150 °C. The heating rate was 2 °C/min and the applied stress was 4 N. The samples had a rectangular shape with dimensions of 50 mm × 13 mm × 2 mm, also prepared in rubber molds. Thermogravimetric (TGA) experiments on the epoxy nanocomposites were performed using an SDT2956 (TA instruments, New Castle, DE, USA) thermobalance under Ar inert gas flow (100 cm^3/min) and at a constant heating rate of 10 °C/min in the temperature range of 25–900 °C.

Oxygen permeability measurements were performed on an Analyzer M8001 (Systech Illinois, Johnsburg, IL, USA) according to the Standard Method (ASTM D3985) [22]. The oxygen transmission rate (OTR, cc/m² day) and permeability (OTR specimen thickness, cc mm/m² day) were measured at a constant temperature of 23 °C and zero relative humidity (0% RH). The specimens tested were the pristine epoxy polymer disks and the epoxy-clay nanocomposite disks, with an average thickness of 2 mm.

2.3. Characterization of Coated Specimens and Protective Properties

Four types of specimens were tested in both salt spray and electrochemical experiments: (a) blank (not coated); (b) coated with pristine glassy epoxy polymer; (c) and (d) coated with the two types of epoxy-clay nanocomposites, i.e., with I.28E and I.30E organoclays. The times of exposure were 1, 2, or 4 days. The salt spray tests were carried out in a corrosive environment of 100% saturated moisture + 5 wt % NaCl solution, according to Standard Methods (ASTM B117) [23,24]. For each tested specimen the weight loss/gain during the exposure in the corrosive environment was measured and the corrosion behaviour was examined by optical and microscopic (scanning electron microscopy) investigation. Three specimens were prepared for each salt spray test. Each tested specimen was washed in clean running water to remove salt deposits from their surface, dried, and weighted, according to ASTM standard B117 [23]. Different specimens were used for the calculation of the corrosion weight loss at 1 or 4 days. After exposure for 1, 2, or 4 days in a corrosive environment of 3.5% NaCl solution, electrochemical impedance spectroscopy measurements were carried out according to Standard Methods (ASTM G106, ASTM B457) [25,26]. Impedance measurements in the controlled potential mode were performed with a system consisting of an impedance spectrum analyzer (Zahner Elektrik IM6, potensiostat DC ±10 V, ±3 A, frequency generator and analyzer, Zahner Elektrik GmbH & Co. KG, Kronach, Germany) connected in serial to a PC terminal computer. We used the THALES evaluation software (Zahner Elektrik GmbH & Co. KG) that runs under the TASC system (Zahner Elektrik GmbH & Co. KG) and combines the MS-DOS system (Microsoft) of the PC terminal with the AMOS/ANDI data acquisition system (Zahner Elektrik GmbH & Co. KG) in a IM6 unit. The frequency scan was carried out over a range from 10,000 Hz (10 kHz) to 0.1 Hz (100 MHz). In all the measurements, ten frequency points per decade were taken and the potential amplitude was 10 mV. A conventional 3 compartment glass cell was used. A platinum foil with a surface area of 2 cm² was used as the counter electrode. A fritted glass separated the anodic compartment. A saturated calomel reference electrode was placed close to the cathode through a Luggin capillary. The supporting electrolyte was 0.1 M LiClO$_4$. All impedance measurements were carried out at 25 °C, in de-aerated conditions and at the potential value of the corrosion potential (E_{corr}). The open circuit potential (OCP) for all types of specimens and times of exposure also was determined.

3. Results and Discussion

3.1. Structure and Morphology of Organo-Clays and Epoxy-Clay Nanocomposites

The XRD patterns of the parent sodium montmorillonite clay (Na$^+$-PGW) and the two organoclays (Nanomer I.28E and I.30E) are shown in Figure 2.

As can be seen from the patterns and the d-spacing data in Figure 2, the hydrophilic Na$^+$-PGW parent clay exhibits a basal spacing of 12.5 Å (corresponding to the interlayer distance of approximately 2.5 Å), due to the presence of water molecules (12 wt % moisture) in the region between the aluminosilicate clay layers (intragallery). The ion-exchange of Na$^+$ cations with octadecylammonium ions in both the organo-clay samples was as high as 93%–95% (determined by carbon analysis), resulting in a significant increase of the basal spacing (23.5–24.5 Å) for both Nanomer I.30E (exchanged with primary onium ions) and Nanomer I.28E (exchanged with quaternary onium ions). The relatively broader and less intense (001) peak in the XRD pattern of the organoclay I.30E compared to that of organoclay I.28E indicates that there was greater disorder of the intercalated layered structure and a broader distribution of basal spacings.

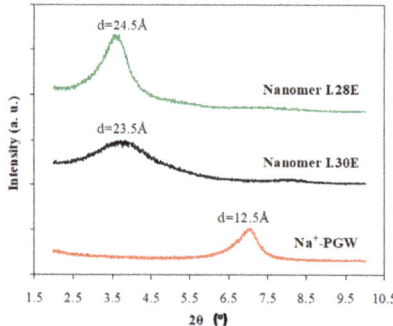

Figure 2. X-ray diffraction (XRD) patterns of the inorganic Na^+-montmorillonite (Na^+-PGW) clay and the organo-montmorillonites Nanomer I.30E and Nanomer I.28E modified by primary and quaternary octadecylammonium ions, respectively.

The structure of epoxy-clay nanocomposites, i.e., the degree of clay nanolayer intercalation or exfoliation within the bulk epoxy polymer, was studied by XRD and HRTEM experiments. The XRD results from the epoxy-(organo) clay nanocomposites are shown in Figure 3.

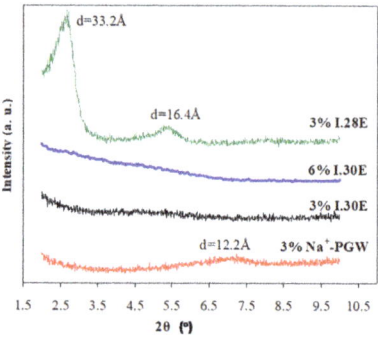

Figure 3. XRD patterns of glassy (EPON 828RS + D-230 Jeffamine) epoxy—clay nanocomposites with inorganic clay Na^+-PGW and the two organoclays I.28E and I.30E; the weight percent of organoclay addition has been estimated on a silicate basis and was 3 and 6 wt %.

The XRD pattern of the epoxy composite prepared with the parent inorganic clay Na^+-PGW (3 wt %) exhibited the characteristic peak (d-spacing of 12.2 Å) of the parent inorganic clay (the epoxy polymer is amorphous), indicating that no intercalation of the epoxy polymer between the clay nanolayers had occurred, as was expected, due to the hydrophilic nature of the inorganic clay surfaces. By comparing the XRD patterns of the nanocomposites prepared by the I.28E (modified with quaternary C18 alkylammonium ions) and I.30E (modified with primary C18 alkylammonium ions) organoclays, we showed that the latter organoclay enabled the formation of an exfoliated clay nanocomposite structure (indicated by the absence of XRD peaks due to ordered nanolayers) in contrast to the former organoclay, which induced the formation of a highly ordered intercalated structure. The actual presence of clay nanolayers in the nanocomposite specimen of I.30E was verified by the XRD peak in the range 60°–65° 2θ, which was attributed to the crystalline structure of the aluminosilicate clay layers (not shown here for brevity). The basal spacing of I.28E increased from 24.5 Å in the powder organoclay to 32.2 Å in the nanocomposite, thus indicating that single or double layers of epoxy polymer chains were intercalated between the organo-functionalized clay nanolayers. The positive effect of the primary alkylammonium ions versus quaternary ions was attributed to the catalytic effect of the acidic protons from the primary alkylammonium ions, which initiate polymerization within the gallery space, thus facilitating exfoliation of the nanolayers and their dispersion in the bulk polymer [18].

The structure of the nanocomposites and the dispersion of the clay layers in the coatings were further studied with high resolution TEM (HRTEM) images. Representative images of the glassy epoxy-clay nanocomposites prepared by 6 wt % organoclay I.30E and 6 wt % organoclay I.28E are shown in Figure 4A–C, respectively.

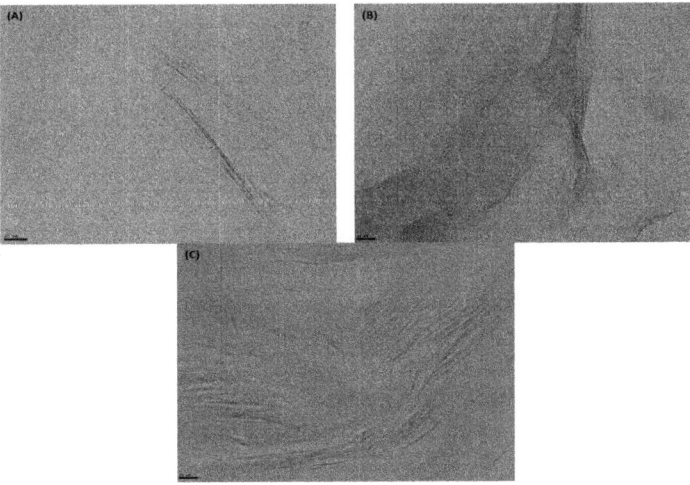

Figure 4. Representative TEM images of partially exfoliated glassy epoxy—clay nanocomposites prepared by 6 wt % I.30E (**A**,**B**) and of intercalated epoxy—clay nanocomposite prepared by 6 wt % I.28E (**C**). Scale bar = 20 nm.

The main difference between the I.30E and I.28E organoclays is that the first is modified with a primary C18 alkylammonium ion while the second is modified with the same alkylammonium ion but in its quaternary form. Although their powder XRD patterns seem very similar due to the successful incorporation of these two onium ions with similar size, when mixed with the epoxy resin a substantial difference in their dispersion has been identified, resulting in two different nanocomposite structures. The primary ammonium ions (I.30E) provide acidic H^+ ions which can catalyse the fast opening of the epoxy rings, thus enhancing the cross-linking polymerization of the epoxy monomers by the amine curing agents. As a result, a partially exfoliated clay nanolayer structure is formed with enhanced interfacial interactions. On the other hand, quaternary ammonium ions are significantly weaker proton donors. So, when in contact with the epoxide, they react with a slower rate thus resulting in an overall slower cross-linking polymerization and eventually leading to intercalated nanocomposite structures.

The above suggested clay dispersion and nanocomposite structures are supported by the XRD patterns, where in the case of I.30E no diffraction peaks can be observed even in the case of the 6 wt % filler (the actual presence of clay in the XRD specimens was verified by the high angle XRD peaks owing to the crystalline structure of the clay). On the other hand, clear distinct XRD peaks are observed in the pattern of the I.28E nanocomposite. From the increase of the d-spacing when compared to the pattern of the I.28E organoclay, it is clear that polymer chains have been inserted within the clay galleries and have induced a further broadening of the interlayer space.

The exfoliated or at least partially exfoliated clay nanolayer structure of the 6 wt % I.30E nanocomposite has been also verified by the TEM images (Figure 4). The presented images are representative of the bulk nanocomposite and show the presence of either isolated single clay nanolayers or bundles of ca. less than 7–8 layers, with varying interlayer distance. It is also known from the literature that, when the distance among the galleries is higher than 50–70 Angstrom and the number of layers is less than 8–10, no diffraction peak can be observed in the XRD. These observations apply also for the 6 wt % I.30E nanocomposite in our study. On the other hand, as can be seen in Figure 4C, the nanocomposite prepared with the organoclay I.28E consists mostly of bundles with more than ca. 20 highly oriented and stratified nanolayers, capable of inducing the XRD profile observed in the respective patterns.

3.2. Properties of Epoxy-Clay Nanocomposites

The mechanical and thermomechanical properties of the pristine epoxy polymer as well as the glassy epoxy nanocomposites are summarized in Table 3.

Table 3. Tensile and thermo-mechanical properties of glassy epoxy–clay nanocomposites.

Sample	Tensile Properties			Dynamic Mechanical Analysis			
	Stress at Break (MPa)	Elongation at Break (%)	Elastic Modulus (MPa)	Storage Modulus at 40 °C (MPa)	Storage Modulus at 100 °C (MPa)	T_g (°C)	
Pristine	62.5	7.9	3026	1660	16	84.5	
3% Na$^+$-PGW	64.9	7.3	3139	1750	19	86.9	
3% I.30E	69.7	6.4	3315	2847	26	82.6	
3% I.28E	57.5	6.0	3030	2540	27	85.5	
6% I.30E	55.8	3.7	3514	2635	34	78.5	
6% I.28E	47.3	3.4	3279	2395	31	81.3	

As can be seen from the data in Table 3, the glassy epoxy nanocomposites prepared by 3 wt % inorganic sodium MMT (Na$^+$-PGW) exhibited slightly improved tensile strength and elastic modulus with a concomitant small decrease of the strain at breaking point. Although the XRD data (Figure 3) showed that the clay nanolayers were not intercalated by the epoxy resin in the (nano)composite sample, it can still be suggested that the observed changes in mechanical properties, with even 3 wt % filler, were derived from the interaction of the epoxy matrix with the external surfaces of the micro-particles of the inorganic clay. The absence of organic modifiers in the inorganic clays allows direct adhesion of the polymer chains to the stiff inorganic external surfaces of the clay particles. In the case of the epoxy nanocomposite prepared using 3 wt % of the I.28E organoclay (modified with quaternary C18 alkylammonium ions), the stress and elongation at breaking load were slightly reduced (the elastic modulus remained unchanged), while with the I.30E organoclay (modified with primary acidic C18 alkylammonium ion), the tensile strength was increased by 12% (with a slight reduction of elongation at the breaking load) and the elastic modulus increased by 10%. In the exfoliated structure of this nanocomposite, the nanolayers were highly dispersed, thus increasing the polymer volume fraction that was being affected by the presence of the stiff clay layers. In addition, the catalytic function of the acidic/primary alkylammonium ions in initiating the epoxy polymerization via the epoxy ring opening led to a stronger interaction of the polymer chains with the surface of the clay layers, compared to the case of the non-reactive quaternary ion modifiers such as those in the I.28E organoclay. An increase in the organoclay concentration from 3 to 6 wt % led to an increase in the elastic modulus with both organoclays (up to a 16% increase compared to the pristine glassy epoxy), with a concomitant decrease in the stress and strain at breaking load, most pronounced in the case of the I.28E organoclay. This behavior was typical for glassy polymer matrices such as the glassy epoxy resins, as increasing the stiffness of the nanocomposite (increase of the modulus) due to the effect of the clay nanolayers leads to more brittle materials that break at lower strain [27,28].

The storage modulus (DMA measurements, Table 3) of the glassy epoxy matrix at 40 °C (glassy region) changed slightly (±5%) with the addition of 3 wt % (silicate basis) of the inorganic sodium MMT, but increased remarkably with the addition of 3 wt % (silicate basis) of both the C18 alkylammonium modified organoclays (70% increase with the I.30E organoclay modified with primary

C18 alkylammonium ions and 50% increase with the I.28E organoclay modified with quaternary C18 alkylammonium ions). Both the organoclays also exhibited substantial improvement of the storage modulus at 100 °C (elastic region) ($\geq$60% increase compared to the pristine polymer). The T_g generally showed a slight fluctuation (up to ± 2 °C) compared to that of the pristine glassy epoxy polymer with the addition of the inorganic and organo-modified clays. The small decrease of T_g in the case of the nanocomposite prepared by the I.30E organoclay can be attributed to the plasticizing effect induced by the chains of C18 alkylammonium ion modifier, which becomes more pronounced in this exfoliated clay nanocomposite compared to the intercalated structures formed with the I.28E organoclay. As mentioned above, in the exfoliated nanocomposite structure a larger volume of the polymer matrix is affected by the organo-modified clay surfaces due to the high dispersion of isolated clay layers. The direct interaction with the stiff inorganic layers increases the modulus of the nanocomposite while, on the other hand, penetration of the modifier's dangling chains within the polymer network reduces the T_g. A greater decrease of the T_g is observed as the organoclay loading increases, due to the increasing concentration of the C18 alkylammonium modifier within the polymer matrix. The storage modulus at 40 °C decreased with the addition of 6 wt % organoclay compared to the nanocomposites with 3 wt % clay, but it was still higher compared to that of the pristine epoxy polymer. The addition of a small percentage of organoclay ($\leq$3 wt %) is improving the thermo-mechanical properties without causing a distortion of the epoxy crosslinked network in the glassy region. A further increase of the organoclay content ($\geq$6 wt %) may cause defects in the polymer network, increase the stiffness substantially, and restrict the flexibility of the polymer chains (glassy region). While the temperature increases, a small degree of elasticity (polymeric chains movement) is induced in the system thus allowing the organoclay sheets to penetrate among the epoxy chains without interrupting the epoxy network (rubbery state). The result of this thermo-mechanical behavior is a small deterioration of the storage modulus values in the glassy region when increasing the organoclay content and in contrast, an increase of these values in the rubbery region.

The TGA curves of the pristine glassy epoxy polymer and the respective epoxy nanocomposites are shown in Figure 5.

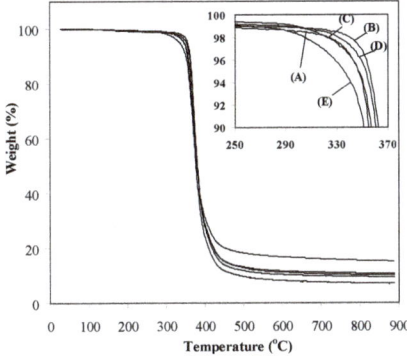

Figure 5. Thermogravimetric analysis (TGA) curves of (A) glassy epoxy (EPON 828RS + D-230 Jeffamine) and epoxy—clay nanocomposites with (B) Na$^+$-PGW, (C) I.30E, (D) I.28E (3 wt %, silicate basis), and (E) I.30E (6 wt %, silicate basis).

From the curves in Figure 5, it can be seen that the presence of the clay filler had a minor effect on the thermal stability of the epoxy polymer, which can be evaluated by comparing the percent weight loss at a certain temperature (for instance at 350 °C, as can be seen in the inset of Figure 5). This behavior can be attributed to the relatively low clay loading, at least for the nanocomposites with 3 wt % clay, which is not sufficient to significantly affect the thermal stability and decomposition rate of the polymeric matrix. The addition of 6 wt % I.30 organoclay induced a slightly faster decomposition, as can be seen in the inset of Figure 5, possibly due to the acidic function of the primary alkylammonium ions of this organoclay, which can catalyze the pyrolysis/decomposition of the polymeric chains.

With regard to the barrier properties of the epoxy polymer and the epoxy-clay nanocomposites, a 30% and 40% reduction of oxygen permeability was observed for the nanocomposites prepared by the addition of I.28E and I.30E organoclays (6 wt %), respectively, compared to that of the pristine polymer (oxygen permeability: 105 cc mm/m^2 day). The reduction of permeability with the addition of the inorganic clay was not significant.

3.3. Salt Spray Corrosion Tests

The results of the weight loss of the steel specimens (bare and covered with pristine epoxy and nanocomposites) upon salt spray exposures are summarized in Figure 6. Optical images of the specimens are shown in Figure 7, while representative scanning electron (SEM) images of the bare and coated specimens are shown in Figure 8.

From the weight loss results (Figure 6) it can be seen that after 1 day of exposure the weight loss of bare steel was about 0.16 wt % and was raised to 1 wt % after 4 days. The neat epoxy polymer showed a relatively good protection efficiency by reducing the weight loss, at 0.07 wt % and 0.4 wt % for the first and fourth day, respectively. A greater improvement in protection was offered by both the epoxy nanocomposites (with I.28E and I.30E organoclays) which inhibited the weight loss, exhibiting similar very low values of ca. 0.03–0.04 wt % and 0.15–0.18 wt % after 1 and 4 days of exposure. These latter values correspond to 75% and 80–90% less weight loss, after 1 and 4 days of exposure, by the use of the epoxy nanocomposites compared to the pristine steel specimens. The differences in corrosion between the bare steel specimen and those covered by the thin epoxy polymer or the epoxy nanocomposites, can be also clearly observed in the respective optical photographs, after 1 and 4 days of exposure, as shown in Figure 7. From these images it follows that the corrosion of coated specimens, especially I.28 and I.30 coated steel—Figure 7c,d, after 1 or 4 days of exposure, is less intense than the corresponding bare steel specimens. The SEM images of the various specimens before the salt spray test (no exposure) and after 1 or 4 days of exposure are shown in Figure 8. In the same figure the cross-section image of I.30 coated steel is shown. In these images the differences in the initial morphology and the corrosion behaviour of the bare and coated steels as well as the coating thickness and the coating-steel interface can be observed.

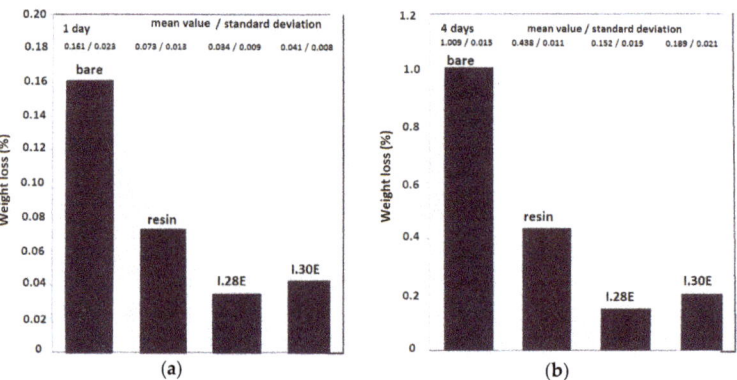

Figure 6. Weight loss results from salt spray experiments for bare steel and steel coated with pristine epoxy resin, and nanocomposites with I.28 and I.30 organo-clays, after 1 day (**a**) and 4 days (**b**) of exposure in a corrosive environment of 100% saturated moisture + 5 wt % NaCl solution.

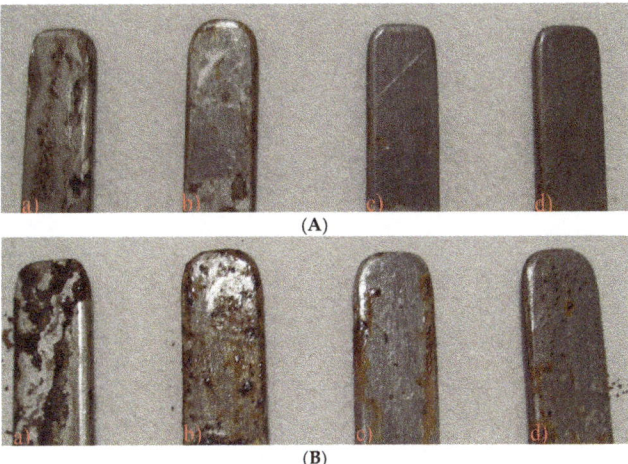

Figure 7. Salt spray corrosion test samples after 1 day (**A**) and 4 days (**B**) of exposure in a corrosive environment of 100% saturated moisture + 5 wt % NaCl solution: (a) bare steel; (b) coated with pristine resin; (c) coated with I.28E nanocomposite; (d) coated with I.30E nanocomposite.

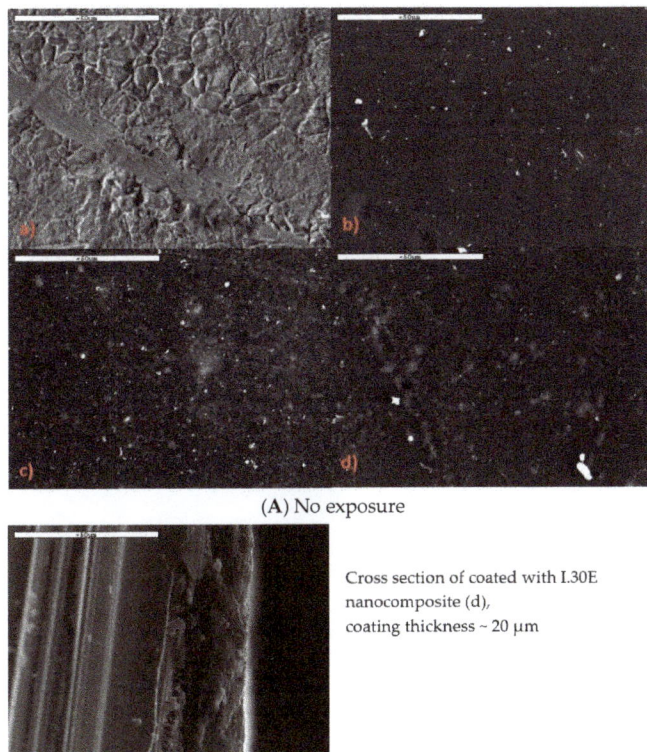

(**A**) No exposure

Cross section of coated with I.30E nanocomposite (d), coating thickness ~ 20 μm

(**B**) Cross section

Figure 8. *Cont.*

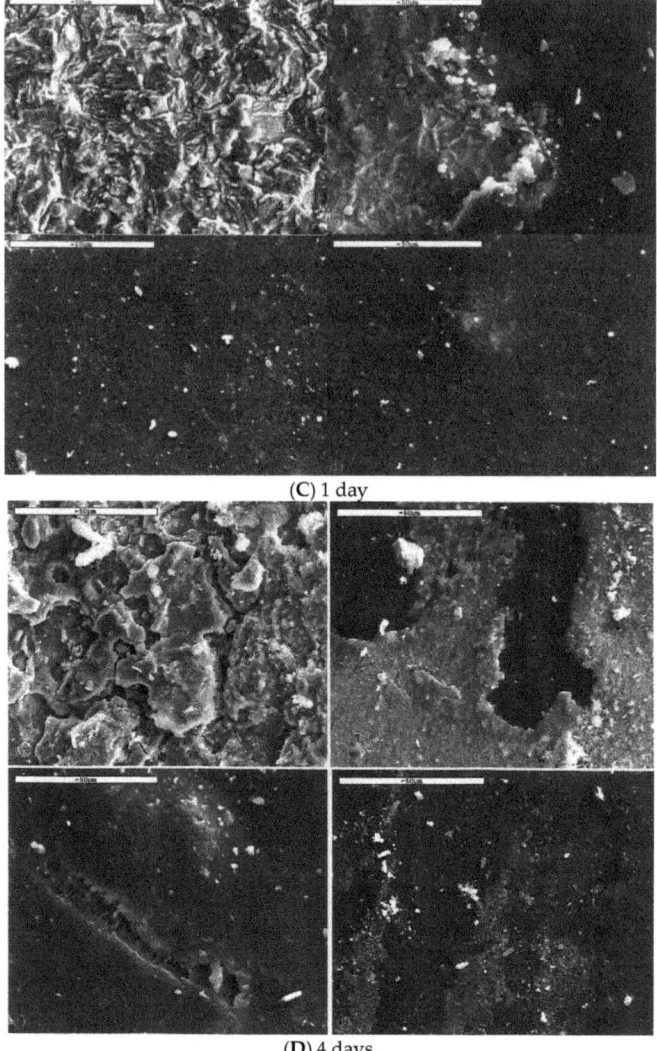

(C) 1 day

(D) 4 days

Figure 8. Scanning electron micrographs of initial (no exposure) (**A**), cross section of steel coated with I.30E (**B**), and salt spray tested specimens after 1 day (**C**) and 4 days (**D**) of exposure in a corrosive environment of 100% saturated moisture + 5 wt % NaCl solution: (a) bare steel; (b) coated with pristine resin; (c) coated with I.28E nanocomposite; (d) coated with I.30E nanocomposite.

3.4. Electrochemical—Open Circuit Potential Measurements

The results of the open circuit potential measurements (OCP) of all types of bare or coated specimens after 1 or 4 days of exposure are shown in Figure 9.

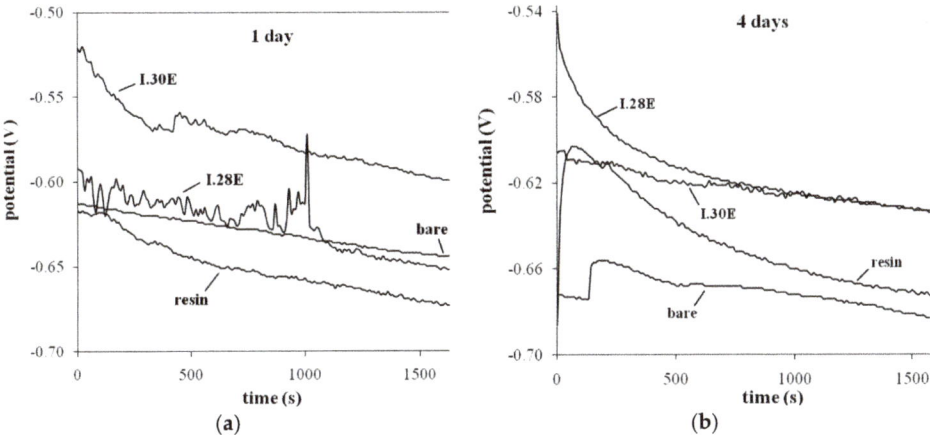

Figure 9. Open circuit potential measurements of all types of bare or coated specimens after 1 day (**a**) and 4 days (**b**) of exposure in a corrosive environment of 3.5 wt % NaCl solution.

Open circuit potential measurements were used for a preliminary and indicative prediction of the corrosion behavior. As it is known, the general criterion for the estimation of corrosion behavior, based on open circuit potential values, is that the more anodic (lower negative value) the OCP, the lower the corrosion susceptibility. However, this criterion is not absolute. The estimation depends on the variations of the potential during the monitoring period and an indication of better corrosion behavior is the stability of the corrosion potential with time [2,16,23,29]. From the open circuit potential with time diagram, after 1 day of exposure (Figure 9), it can be seen that the more anodic value of OCP was obtained for the I.30E nanocomposite coating (−595 mV) and the corresponding more cathodic value was obtained for bare steel (−670 mV). All curves moved linearly towards the cathodic direction and oscillations were observed mainly in the case of the I.28E nanocomposite coating. This decrease of OCP indicated a continuous dissolution of the surface layers of the steel specimens. From the open circuit potential with time diagram, after 4 days of exposure (Figure 9), it can be seen that the more anodic value of OCP was obtained for both the I.30 and I.28 nanocomposite coatings (−625 mV) and the corresponding more cathodic value was obtained for bare steel (−680 mV). OCP curves of bare and resin coated specimens move initially in the anodic direction and after that continuously towards more cathodic values, indicating the formation of a non-passive oxide layer (in the case of the bare specimen) or the non-passive behavior of the coating (in the case of the resin coated specimen) that cracks and so the potential decreases and corrosion increases. The I.30 nanocomposite coating curve only decreased slightly (from −605 to −625 mV), indicating an increased stability of the coating and better corrosion behavior.

The electrochemical impedance response is a fundamental characteristic of an electrochemical system. Knowledge of the frequency dependency of impedance for a corroding system enables a model equivalent electrical circuit describing that system to be created [29–32]. Processing of the experimental data and fitting of the electrochemical impedance measurements was based on the equivalent circuit shown in Figure 10 for bare and coated steels, using a non-linear regression analysis. The results are presented in Figures 11–16 (Nyquist diagrams) and Table 4 (Resistance values).

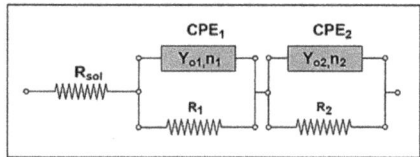

Figure 10. Electrical equivalent circuit model simulating a corroding system metal/coating/electrolyte, R_s = solution resistance, R_1, R_2 = ohmic resistances, CPE_1, CPE_2 = constant phase elements with admittance $Y_0(j\omega)^n$.

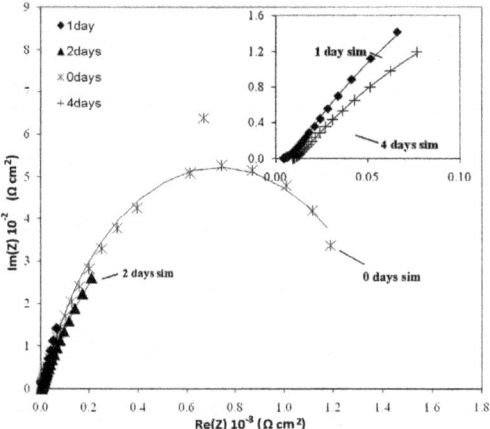

Figure 11. Nyquist plots of impedance spectra, $R_e(Z)$, real part of impedance, and $I_m(Z)$, imaginary part of impedance, for bare specimens after 0, 1, 2, or 4 days of exposure in a corrosive environment of 3.5 wt % NaCl solution.

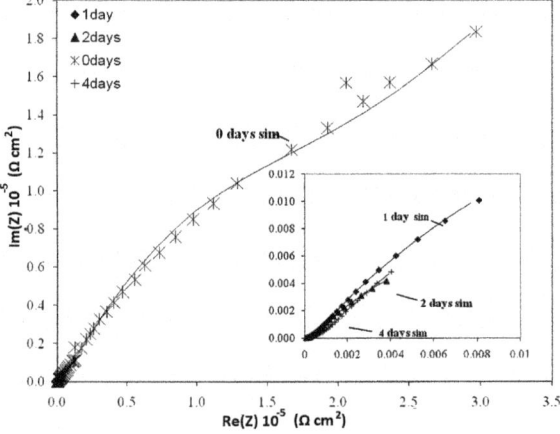

Figure 12. Nyquist plots of impedance spectra, $R_e(Z)$, real part of impedance, and $I_m(Z)$, imaginary part of impedance, for epoxy coated specimens after 0, 1, 2, or 4 days of exposure in a corrosive environment of 3.5 wt % NaCl solution.

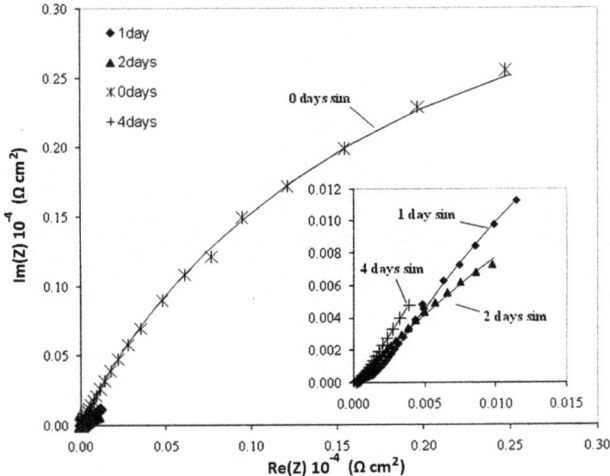

Figure 13. Nyquist plots of impedance spectra, R_e (Z), real part of impedance, and I_m (Z), imaginary part of impedance, for I.28 nanocomposite coated specimens after 0, 1, 2, or 4 days of exposure in a corrosive environment of 3.5 wt % NaCl solution.

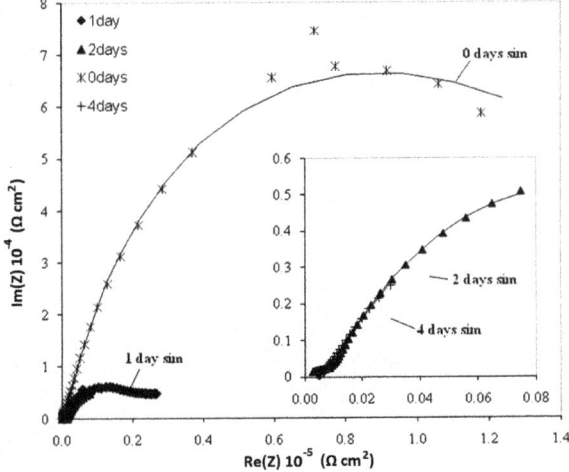

Figure 14. Nyquist plots of impedance spectra, R_e (Z), real part of impedance, and I_m (Z), imaginary part of impedance, for I.30 nanocomposite coated specimens after 0, 1, 2, or 4 days of exposure in a corrosive environment of 3.5 wt % NaCl solution.

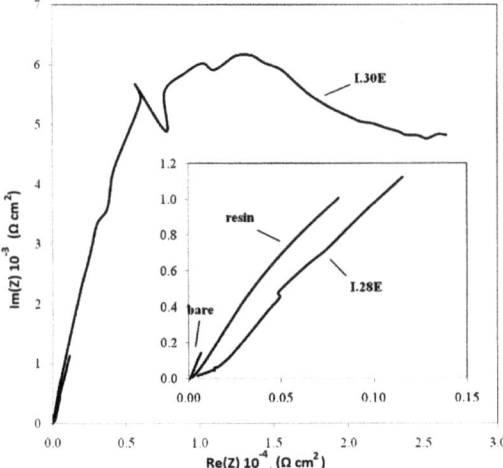

Figure 15. Nyquist plots of impedance spectra, R_e (Z), real part of impedance, and I_m (Z), imaginary part of impedance, for bare and coated specimens after 1 day of exposure in a corrosive environment of 3.5 wt % NaCl solution.

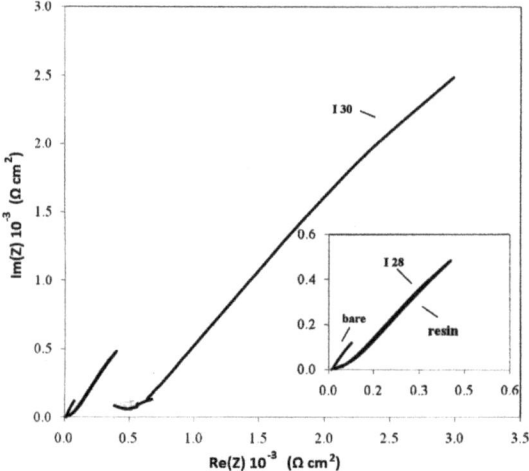

Figure 16. Nyquist plots of impedance spectra, R_e (Z), real part of impedance, and I_m (Z), imaginary part of impedance, for bare and coated specimens after 4 days of exposure in a corrosive environment of 3.5 wt % NaCl solution.

Table 4. Resistance values (Ω cm^2) for bare and coated specimens.

Parameters of Equivalent Circuit	Bare—0 day	Bare—1 day	Bare—2 days	Bare—4 days	Epoxy—0 day	Epoxy—1 day	Epoxy—2 days	Epoxy—4 days
R_{sol}	1.27	4.00	2.22	9.08	3.14×10^2	2.00	4.00	1.62×10
R_1	1.42×10^3	9.93×10^2	6.55×10^2	3.67	2.15×10^3	5.43×10	1.87×10	5.27×10^3
Y_{01}	3.93×10^{-4}	1.08×10^{-2}	3.97×10^{-2}	3.51×10^{-2}	1.62×10^{-6}	4.43×10^{-3}	1.05×10^{-3}	2.33×10^{-3}
n_1	8.08×10^{-1}	8.72×10^{-1}	7.01×10^{-1}	4.79×10^{-1}	6.26×10^{-1}	2.93×10^{-1}	4.86×10^{-1}	6.76×10^{-1}
R_2	2.48×10	-	-	1.03×10^2	6.78×10^5	7.26×10^3	5.38×10^3	7.21×10^1
Y_{02}	1.26×10^{-3}	5.92×10^{-2}	4.22×10^{-2}	9.99×10^{-2}	2.10×10^{-6}	1.03×10^{-3}	2.31×10^{-3}	1.34×10^{-3}
n_2	7.10×10^{-1}	3.37×10^{-1}	3.17×10^{-1}	7.61×10^{-1}	5.68×10^{-1}	6.93×10^{-1}	6.19×10^{-1}	4.60×10^{-1}
R_{tot} ($R_1 + R_2$)	1.44×10^3	9.93×10^2	6.55×10^2	1.03×10^2	6.80×10^5	7.31×10^3	5.40×10^3	5.34×10^3

Parameters of Equivalent Circuit	I.28E—0 day	I.28E—1 day	I.28E—2 days	I.28E—4 days	I.30E—0 day	I.30E—1 day	I.30E—2 days	I.30E—4 days
R_{sol}	1.38×10	1.37×10	1.51×10	1.40×10	7.67×10	9.66	1.11×10	1.50×10
R_1	9.36	2.12×10^2	8.12×10^3	7.35×10^3	4.93×10^4	1.80×10^4	1.09×10^3	6.03×10^2
Y_{01}	1.37×10^{-4}	7.18×10^{-4}	8.98×10^{-4}	2.33×10^{-3}	9.74×10^{-6}	2.63×10^{-6}	4.08×10^{-5}	4.26×10^{-5}
n_1	1.00	3.15×10^{-1}	4.94×10^{-1}	6.38×10^{-1}	1.00	6.53×10^{-1}	3.79×10^{-1}	2.96×10^{-1}
R_2	7.99×10^4	8.99×10^3	6.24	4.81×10	1.54×10^5	2.63×10^4	3.02×10^4	2.90×10^4
Y_{02}	3.02×10^{-5}	8.25×10^{-4}	1.63×10^{-4}	1.44×10^{-3}	8.72×10^{-6}	7.49×10^{-5}	1.15×10^{-4}	3.59×10^{-4}
n_2	7.69×10^{-1}	6.41×10^{-1}	7.01×10^{-1}	4.38×10^{-1}	7.10×10^{-1}	4.31×10^{-1}	6.95×10^{-1}	5.79×10^{-1}
R_{tot} ($R_1 + R_2$)	7.99×10^4	9.21×10^3	8.12×10^3	7.40×10^3	2.03×10^5	4.43×10^4	3.13×10^4	2.96×10^4

From the results of the electrochemical impedance measurements (Figures 11–16 and Table 4), it follows that all the coatings used increased the total resistance and both nanocomposites had greater total resistance values than the pristine glassy epoxy polymer, indicating improved protection properties. Thus, the total resistance value, R_{tot}, after 4 days exposure in the corrosive environment, increased from 1.03×10^2 of bare steel, to 5.34×10^3 in the case of pristine resin, to 7.40×10^3 in the case of I.28E, and to 2.96×10^4 (Ω cm^2) in the case of I.30E coated specimens. In the case of bare steel, the total resistance values decrease continuously with exposure time and the relation between these two factors was observed to be linear, indicating a constant corrosion rate. A high total resistance value in the case of pristine resin coated steel in conditions of no prior exposure in the corrosive environment (0 days) was observed, 6.80×10^5 Ω cm^2, that decreased quickly in any conditions of exposure, 5.34×10^3 Ω cm^2 after 4 days, indicating high corrosion rate values and the fast evolution of corrosion. In the case of nanocomposite coated steel, the total resistance values decreased after four days of exposure, from 7.99×10^4 to 7.40×10^3 Ω cm^2 in the case of I.28E and from 2.03×10^5 to 2.96×10^4 Ω cm^2 in the case of I.30E, as it was expected due to corrosion initiation. However, the decrease is much smaller and so the corrosion evolution is also lower.

The superior protection performance, indicated from the higher total resistance value, of the epoxy nanocomposite with the I.30E organo-clay compared to that with the I.28E organic-clay, can be attributed to the higher dispersion of the clay nanolayers in the former nanocomposite (XRD and TEM results), which also induced slightly improved barrier properties.

4. Conclusions

The mechanical, thermomechanical, and barrier properties of all the epoxy—organoclay nanocomposites were improved compared to those of the pristine epoxy polymer. Both the pristine epoxy and the epoxy nanocomposite coatings offered substantial protection to steel from corrosion. The protective properties of the nanocomposite coatings were superior compared to those of the pristine epoxy polymer, as it was revealed from the weight loss results, the optical and microscopy examination of the specimens after the exposure in the corrosive environment, the open circuit potential measurements, and the electrochemical impedance spectroscopy measurements. The protective properties of the nanocomposite coatings varied with the organo-clay used. The epoxy—montmorillonite clay modified with primary octadecylammonium ions, Nanomer I.30E, had a better behaviour than the clay modified with quaternary octadecylammonium ions, Nanomer I.28E. This was attributed to the higher dispersion of the nanolayers in the nanocomposite formed with the I.30E organoclay compared to that formed with I.28E. The enhanced mechanical properties and thermal stability of both epoxy—clay nanocomposites, in combination with their high protection efficiency, renders them as attractive candidates for various demanding coating applications.

Author Contributions: Panagiotis Giannakoudakis, Konstantinos Triantafyllidis and Panagiotis Spathis conceived and designed the experiments; Domna Merachtsaki and Panagiotis Xidas performed the experiments; Domna Merachtsaki, and Panagiotis Xidas analyzed the data; Panagiotis Giannakoudakis, Konstantinos Triantafyllidis and Panagiotis Spathis contributed reagents/materials/analysis tools; Domna Merachtsaki, Panagiotis Xidas and Panagiotis Spathis wrote the paper.

Conflicts of Interest: The authors declare no conflict of interest.

References

1. Hang, T.T.X.; Truc, T.A.; Olivier, M-G.; Vandermiers, C.; Guérit, N.; Pébère, N. Corrosion protection of carbon steel by an epoxy resin containing organically modified clay. *Surf. Coat. Technol.* **2007**, *201*, 7408–7415. [CrossRef]
2. Yeh, J.M.; Huang, H.Y.; Chen, C.L.; Su, W.F.; Yu, Y.H. Siloxane-modified epoxy resin—Clay nanocomposite coatings with advanced anticorrosive properties prepared by a solution dispersion approach. *Surf. Coat. Technol.* **2006**, *200*, 2753–2763. [CrossRef]
3. Kouloumbi, N.; Ghivalos, L.G.; Pantazopoulou, P. Determination of the performance of epoxy coatings containing feldspars filler. *Pigment Resin Technol.* **2005**, *34*, 148–153. [CrossRef]
4. Triantafyllidis, K.S.; LeBaron, P.C.; Park, I.; Pinnavaia, T.J. Epoxy-clay fabric film composites with unprecedented oxygen-barrier properties. *Chem. Mater.* **2006**, *18*, 4393–4398. [CrossRef]
5. Pinnavaia, T.G.; Beall, G.W. (Eds.) *Polymer–Clay Nanocomposites, Wiley Series in Polymer Science*; John Wiley & Sons: Oxford, UK, 2000.
6. Vaia, R.A.; Giannelis, E.P. Polymer Nanocomposites: Status and Opportunities. *MRS Bull.* **2001**, *26*, 394–401. [CrossRef]
7. Allie, L.; Thorn, J.; Aglan, H. Evaluation of nanosilicate filled poly (vinyl chloride-co-vinyl acetate) and epoxy coatings. *Corros. Sci.* **2008**, *50*, 2189–2196. [CrossRef]
8. Dai, C.-F.; Li, P.-R.; Yeh, J.M. Comparative studies for the effect of intercalating agent on the physical properties of epoxy resin-clay based nanocomposite materials. *Eur. Polym. J.* **2008**, *44*, 2439–2447. [CrossRef]
9. Yu, H.J.; Wang, L.; Shi, Q.; Jiang, G.H.; Zhao, Z.R.; Dong, X.C. Study on nano-$CaCO_3$ modified epoxy powder coatings. *Prog. Org. Coat.* **2006**, *55*, 296–300. [CrossRef]
10. Hosseini, M.G.; Raghibi-Boroujeni, M.; Ahadzadeh, I.; Najjar, R.; Seyed Dorraji, M.S. Effect of polypyrrole—Montmorillonite nanocomposites powder addition on corrosion performance of epoxy coatings on Al 5000. *Prog. Org. Coat.* **2009**, *66*, 321–327. [CrossRef]
11. Rezaul Karim, M.; Hyun Yeum, J. In situ intercalative polymerization of conducting polypyrrole/montmorillonite nanocomposites. *J. Polym. Sci. Part B* **2008**, *46*, 2279–2285. [CrossRef]
12. Pinnavaia, T.J. Intercalated Clay Catalysts. *Science* **1983**, *220*, 365–371. [CrossRef] [PubMed]
13. Triantafillidis, C.S.; LeBaron, P.C.; Pinnavaia, T.J. Thermoset Epoxy—Clay Nanocomposites: The Dual Role of α, ω-Diamines as Clay Surface Modifiers and Polymer Curing Agents. *J. Solid State Chem.* **2002**, *167*, 354–362. [CrossRef]
14. Triantafillidis, C.S.; LeBaron, P.C.; Pinnavaia, T.J. Homostructured mixed inorganic-organic ion clays: A new approach to epoxy polymer-exfoliated clay nanocomposites with a reduced organic modifier content. *Chem. Mater.* **2002**, *14*, 4088–4095. [CrossRef]
15. Kouloumbi, N.; Moundoulas, P. Anticorrosive performance of organic coatings on steel surfaces exposed to deionized water. *Pigment Resin Technol.* **2002**, *4*, 206–215. [CrossRef]
16. Yeh, J.-M.; Hsieh, C.-F.; Jaw, J.-H.; Kuo, T.-H.; Huang, H.-Y.; Lin, C.-L.; Hsu, M.-Y. Organo-Soluble Polyimde (ODA-BSAA)/Montmorillonite Nanocomposite Materials Prepared by Solution Dispersion Technique. *J. Appl. Polym. Sci.* **2005**, *95*, 1082–1090. [CrossRef]
17. Sung, J.H.; Choi, H.J. Effect of pH on physical characteristics of conducting poly (O-ethoxyaniline) nanocomposites. *J. Macromol. Sci. Part B* **2005**, *44*, 365–375. [CrossRef]

18. Xidas, P.I.; Triantafyllidis, K.S. Effect of the type of alkylammonium ion clay modifier on the structure and thermal/mechanical properties of glassy and rubbery epoxy-clay nanocomposites. *Eur. Polym. J.* **2010**, *46*, 404–417. [CrossRef]
19. Triantafyllidis, K.S.; Xidas, P.I.; Pinnavaia, T.J. Alternative Synthetic Routes to Epoxy Polymer—Clay Nanocomposites using Organic or Mixed-Ion Clays Modified by Protonated Di/Triamines (Jeffamines). *Macromol. Symp.* **2008**, *267*, 41–46. [CrossRef]
20. *ASTM A366/A366M-97e1 Standard Specification for Commercial Steel (CS) Sheet, Carbon (0.15 Maximum Percent) Cold-Rolled (Withdrawn 2000)*; ASTM International: West Conshohocken, PA, USA, 1998.
21. *ASTM D638-14 Standard Test Method for Tensile Properties of Plastics*; ASTM International: West Conshohocken, PA, USA, 2014.
22. *ASTM D3985-05(2010)e1 Standard Test Method for Oxygen Gas Transmission Rate Through Plastic Film and Sheeting Using a Coulometric Sensor*; ASTM International: West Conshohocken, PA, USA, 2010.
23. *ASTM B117 Standard Practice for Operating Salt Spray (Fog) Apparatus*; ASTM International: West Conshohocken, PA, USA, 2003.
24. Baboian, R. *Corrosion Tests and Standards: Application and Interpretation*, 2nd ed.; ASTM International: West Conshohocken, PA, USA, 2005.
25. *ASTM G106 Standard Practice for Verification of Algorithm and Equipment for Electrochemical Impedance Measurements*; ASTM International: West Conshohocken, PA, USA, 2004.
26. *ASTM B457 Standard Test Method for Measurement of Impedance of Anodic Coatings on Aluminum*; ASTM International: West Conshohocken, PA, USA, 2003.
27. Becker, O.; Simon, G.P.; Dusek, K. *Inorganic Polymeric Nanocomposites and Membranes*; Springer: Berlin, Germany, 2005.
28. Ho, M.-W.; Lam, C.-K.; Lau, K.-T.; Ng, D.H.L.; Hui, D. Mechanical properties of epoxy-based composites using nanoclays. *Compos. Struct.* **2006**, *75*, 415–421. [CrossRef]
29. Bard, A.J.; Faulkner, L.R. *Electrochemical Methods: Fundamentals and Applications*, 2nd ed.; John Wiley & Sons: Oxford, UK, 2001.
30. Orazem, M.; Tribollet, B. *Electrochemical Impedance Spectroscopy*; the Electrochemical Society Series; John Wiley & Sons: Oxford, UK, 2008.
31. Perez, N. *Electrochemistry and Corrosion Science*; Kluwer Academic Publishers: New York, NY, USA, 2004.
32. Brett, C.M.A.; Brett, A.M.O. *Electrochemistry Principles, Methods, and Applications*; Oxford University Press: Oxford, UK, 1994.

 © 2017 by the authors. Licensee MDPI, Basel, Switzerland. This article is an open access article distributed under the terms and conditions of the Creative Commons Attribution (CC BY) license (http://creativecommons.org/licenses/by/4.0/).

Review

Emerging Corrosion Inhibitors for Interfacial Coating

Mona Taghavikish [1], Naba Kumar Dutta [1,2,*] and Namita Roy Choudhury [1,2,*]

1 Future Industries Institute, University of South Australia, Mawson Lakes Campus, Mawson Lakes, Adelaide, SA 5095, Australia; mona.taghavikish@mymail.unisa.edu.au
2 School of Chemical Engineering, University of Adelaide, Adelaide, SA 5005, Australia
* Correspondence: naba.dutta@adelaide.edu.au (N.K.D.) or naba.dutta@unisa.edu.au (N.K.D.); Namita.roychoudhury@adelaide.edu.au (N.R.C.) or namita.choudhury@unisa.edu.au (N.R.C.); Tel.: +61-8830-23546 (N.K.D.); +61-8830-23719 (N.R.C.)

Academic Editor: Tony Hughes
Received: 5 September 2017; Accepted: 2 October 2017; Published: 1 December 2017

Abstract: Corrosion is a deterioration of a metal due to reaction with environment. The use of corrosion inhibitors is one of the most effective ways of protecting metal surfaces against corrosion. Their effectiveness is related to the chemical composition, their molecular structures and affinities for adsorption on the metal surface. This review focuses on the potential of ionic liquid, polyionic liquid (PIL) and graphene as promising corrosion inhibitors in emerging coatings due to their remarkable properties and various embedment or fabrication strategies. The review begins with a precise description of the synthesis, characterization and structure-property-performance relationship of such inhibitors for anti-corrosion coatings. It establishes a platform for the formation of new generation of PIL based coatings and shows that PIL corrosion inhibitors with various heteroatoms in different form can be employed for corrosion protection with higher barrier properties and protection of metal surface. However, such study is still in its infancy and there is significant scope to further develop new structures of PIL based corrosion inhibitors and coatings and study their behaviour in protection of metals. Besides, it is identified that the combination of ionic liquid, PIL and graphene could possibly contribute to the development of the ultimate corrosion inhibitor based coating.

Keywords: ionic liquid; polyionic liquid; graphene; hybrid coating

---

1. Introduction

Corrosion of metal is a significant problem, costing worldwide industries more than $300 billion annually. The inhibitors minimize the rate of corrosion by forming a thin adsorbed film on metal. In the last decades, much attention has been focused on the need to design and develop new and emerging materials for corrosion protection. As an example, nanomaterials, biomaterials, corrosion inhibitors, sol-gel coatings, self-healing and smart materials. Out of these, self-healing coating and corrosion inhibitors are an emerging and broad field.

A self-healing system is inspired from biological systems that have inherent ability to repair damage via healing mechanisms and is categorized into three types; namely, microencapsulation, vascular based and intrinsic materials [1]. These systems have the ability to repair the damage caused due to mechanical stress or energy and to recover their functionality using resources inherently available to them. On the other hand, corrosion can be inhibited or controlled by introducing a stable protective layer of inert metals, conductive polymers, inorganic compound or monolayers of graphitic or heterocyclic structure between a metal and a corrosive environment. A corrosion inhibitor is a chemical constituent which, when added in small amount to the metal environment, diminishes or controls and prevents corrosion. In the oil and chemical industry, inhibitors are considered as the first line of defense against corrosion.

In simpler words, corrosion can be defined as failing of materials by chemical process. Among them the most significant is electrochemical corrosion of metals, in which oxidation process ($M \rightarrow M^{n+} + ne^-$) is helped by the presence of a suitable electron acceptor, sometimes referred to in corrosion science as depolarizer. In general, corrosion is a two-step electrochemical process having both anodic and cathodic sites, with flow of charges (electrons and ions), it is conventional in both wet and dry conditions. Wet corrosion is a major problem to tackle; it is a dominating corrosion at or near room temperature and in presence of an electrolyte, or even in presence of water.

Since corrosion process is a surface reaction, addition of corrosion inhibitor in very small concentration to an interfacial layer can prevent or reduce the corrosion rate of a metal exposed in aggressive environment. Generally there are three mechanisms of the corrosion inhibition as given below [2,3]:

- Adsorption: the inhibitor is chemically adsorbed on the surface of the metal and forms a protective thin film with inhibitor effect.
- Surface layer: formation of an oxide film for protection of the metal surface.
- Passivation: the inhibitor reacts with corrosive elements of aqueous media, forming protective precipitates.

Based on the above mechanism of corrosion inhibitors, they can be classified to three different types; cathodic, anodic, and mixed or adsorption type inhibitors. Corrosion inhibitors that cause the delay in the cathodic reaction are known as cathodic inhibitors. Similarly, the anodic inhibitors slow down the anodic reaction. Those inhibitors that affect both the cathodic and the anodic reactions are known as mixed inhibitors, and these inhibitors generally work by an adsorption mechanism and known as adsorption inhibitors. In general, inorganic inhibitors have either cathodic or anodic actions, while organic inhibitors have both cathodic and anodic actions [2] (Figure 1).

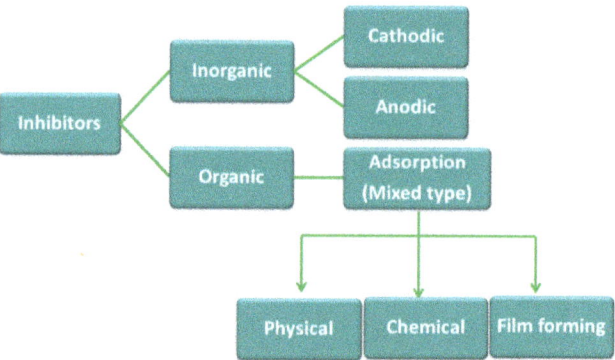

Figure 1. Classification of corrosion inhibitors [2].

Due to the toxicity of inorganic inhibitors, a variety of organic compounds have been used as corrosion inhibitors for the protection of steel specifically in acid medium [4]. In general, organic corrosion inhibitors are more effective than inorganic compounds for protection of steels in acid media. Organic inhibitors work by an adsorption mechanism in which the adsorption of the inhibitor molecule at the metal-solution interface results in formation of a film of inhibitor molecules to protect the surface from the corrosive environment either by physically blocking or by delaying the electrochemical processes [5]. Organic inhibitors generally contain heteroatoms (S, O, or N) and their efficiency is related to the presence of these atoms in the molecule as well as heterocyclic compounds and π electrons [6,7]. This is due to the fact that O, N, and S are found to have higher basicity and electron density and are the key active centres for the adsorption process on the metal surface.

Adsorption inhibitors protect the metal following three possible ways: (1) physical adsorption, (2) chemical adsorption and (3) film formation (Figure 1). Physical (or electrostatic) adsorption is a result of electrostatic attraction between the inhibitor and the metal surface. Physically adsorbed inhibitors interact rapidly, but they are also easily removed from the surface. The most effective inhibitors are those that chemically adsorbed (chemisorbed). Chemisorption occurs as a result of charge sharing or charge transfer between the inhibitor molecules and the metal surface. However, chemisorption is slower than physical adsorption process and is not completely reversible [8].

Film formation mechanism is based on the surface reactions of inhibitor molecules and formation of thin film on the surface with blocking both anodic and cathodic areas. Organic inhibitors are able to form a protective hydrophobic film, adsorbed on the metal surface. In fact, the polar group of the organic molecule is directly attached to metal and the nonpolar end is oriented in a vertical direction to the metal surface. Thus, they can prevent diffusion of corrosive species and establish a barrier against chemical and electrochemical attack [6].

Most organic inhibitors contain at least one functional group. The strength of adsorption of organic inhibitors relies on the charge of this group rather on the hetero atom present in the organic molecule. The structure of the rest of the molecule influences the charge density on the functional group [6]. Most common organic inhibitors belonging to different chemical families such as fatty amides [9,10], pyridines [11,12], imidazolines [13,14], other 1,3-azoles [15,16] and polymers [17] have showed excellent performance as corrosion inhibitors.

While a variety of different inorganic and organic compounds can be used as inhibitors, however, the practical application of many of those inhibitors poses risk for environmental protection standards, cost and toxicity. Thus, there is a strong need to develop efficient and environmentally friendly corrosion inhibitors. Among various classes of compounds, ionic liquids (ILs) have attracted considerable attention in recent years as "green material," because of their attractive properties such as chemical and thermal stability, nonflammability, very low or negligible vapour pressure, high ionic conductivity, a wide electrochemical potential window. They can be used as potential inhibitors whose specific interactions with metal can be tailored through choice of their amphiphilic structures or using them in various forms such as microcapsule, gel, emulsion, nanoparticles or using them in synergistic combinations. Due to high sensitivity of the metal-IL interactions, careful design and tailoring of ionic liquid materials play a crucial role for successful corrosion inhibition application.

Another promising material, graphene a single-atom-thick sheet [18], a flat monolayer of carbon atoms tightly packed into hexagonal honey comb lattice in which carbon atom is sp^2 hybridised has been identified as a next generation inhibitor material for shielding of metal from corrosion as it possesses matchless properties such as excellent thermal and chemical stability, high strength, chemical inertness, permeability to molecules and gases, extremely high aspect ratio, high theoretical specific surface area. From the point of permeability, the hexagonal network of carbon atoms in graphene is so dense that no known material can penetrate through it. However, there are a number of critical challenges related to application of graphene on various metals, which needs significant attention. To date, graphene coatings on metals have been employed using chemical vapour deposition (CVD) or transfer techniques involving high energy consumption, special expensive tools, high temperatures, careful treatments and multistep processes. Such techniques (CVD or transfer) are cumbersome, uneconomical and not very practical for large scale application. On the other hand, there are significant advantages if graphene can be deposited from preformed ink, which is reproducible, and can be used to coat objects of any dimensions as in conventional paints or coatings.

Thus, the key focus of this review is to present and discuss some important results on the physico-chemical properties of the emerging corrosion inhibitors based on IL and graphene to advanced coating applications. Before the presentation of these results, a precise description of the synthesis, characterisation, structure-properties relationship and performance of IL and graphene based inhibitors suitable for anticorrosion applications is given below.

2. Ionic Liquid (IL) Based Corrosion Inhibitors

Among all the different types of synthetic materials, a new class of low toxicity organic compounds known as Ionic liquids (ILs) deserves particular attention due to their rapid growth in a number of applications, they have shown effective performance as inhibitors for various metals and alloys [19,20]. In this section, an insight into ionic liquids will be discussed in detail.

ILs are the low-melting organic salts that are composed of cations and anions that melt below 100 °C [21,22]. The first IL was investigated in 1914 by Paul Walden with his observation on ethylammonium nitrate ([EtNH$_3$][NO$_3$]) with very low melting point of 13–14 °C [21]. Due to the unique properties such as low toxicity, negligible vapor pressure, high thermochemical and electrochemical stabilities, non-flammability, and their ability to act as catalyst, ILs have been used in a large number of applications as an eco-friendly alternative to substitute volatile organic solvents including catalysis [23], separation processes [24,25], analytics [26], lubricants [27], and electrochemical applications [28]. Common ionic liquids are formed by an organic cation (i.e., ammonium, imidazolium, pyridinium, pyrrolidinium, phosphonium, sulfonium) in combination with a complex anion (Scheme 1) [29].

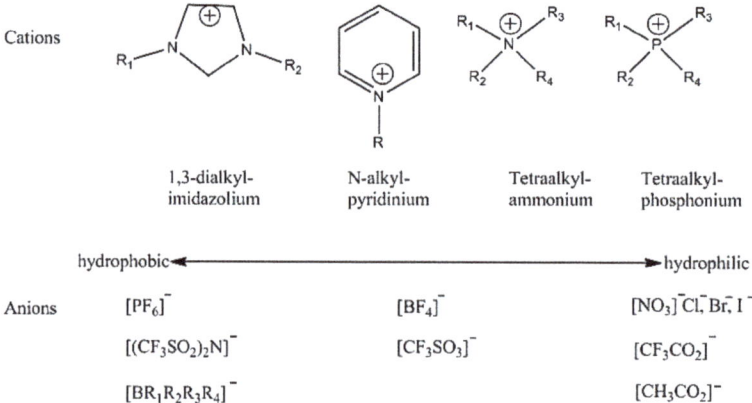

Scheme 1. Most commonly used cations and anions in various ILs.

The common configuration of ILs consists of an amphiphilic group with a long chain, hydrophobic tail, and a hydrophilic polar head [30]. Therefore, due to their molecular configuration, they are able to form micelles and lowering interfacial tension of aggressive media, resulting in an enhancement in surface wetting and adsorption [30–32]. These properties of ILs have a useful effect on surface exposure and may be responsible for the corrosion inhibition of metals. ILs compounds are reported to show corrosion resistant behavior on copper, mild steel and aluminum. Here some of the literature examples will be discussed.

Espinosa et al. [31] studied the corrosion rate and surface interaction of oxygen-free high conductivity (OFHC) copper with two protic ammonium ionic liquids and four aprotic imidazolium species in order to investigate the best candidate for lubricant applications or as precursors of surface coatings. The protic ILs, with no heteroatoms in their composition, are the triprotic di[(2-hydroxyethyl) ammonium] succinate (MSu) and the diprotic di[bis-(2-hydroxyethyl)ammonium] adipate (DAd). The four aprotic ILs contain imidazolium cations with short or long alkyl chain substituents and reactive anions: 1-ethyl-3-methylimidazolium phosphonate ([EMIM]EtPO$_3$H); 1-ethyl-3-methylimidazolium octylsulfate ([EMIM]C$_8$H$_{17}$SO$_4$); 1-hexyl-3-methylimidazolium tetrafluoroborate ([HMIM]BF$_4$) and 1-hexyl-3-methylimidazolium hexafluorophosphate ([HMIM]PF$_6$). As it has been depicted in summary of results in Figure 2, the lowest corrosion rate is observed for the DAd, which gives low mass (Δm)

and surface roughness changes (ΔS_a) and forms adsorbed layers on copper, while MSu forms a dark blue corrosion product by reaction with copper.

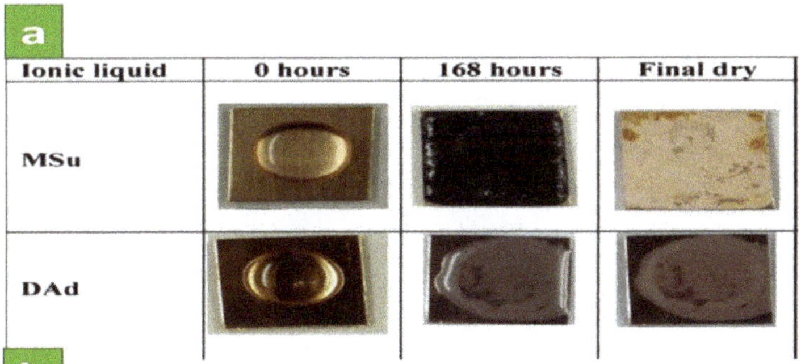

a

Ionic liquid	0 hours	168 hours	Final dry
MSu			
DAd			

b

Ionic liquid	Contact angle (°) (standard deviation)	Δm (%)	ΔS_a (%)
MSu	31.75 (1.19)	−1.40	225
Dad	62.67 (2.19)	−0.04	−3.0

Figure 2. (a) The OFHC copper change during the tests; (b) Contact angles, mass (Δm) and surface roughness (ΔS_a) changes of MSu and DAd after 168 h (Reprinted with permission from [31]. Copyright 2013 Elsevier).

Results show that DAd IL remains colourless during the corrosion tests (Figure 2a) and no precipitates are formed on the copper surface, while MSu forms a dark blue corrosion product that completely covers the copper surface at the end of the test. SEM observation showed more roughness in the case of use of MSu and the presence of oxygen and carbon peaks in EDX analysis. Nevertheless, EDX analysis of DAd shows only the presence of copper. They concluded that the presence of proton donor and acceptor sites in the DAd molecules can form a hydrogen bonded network which as a result will improve their lubricating performance. Moreover, all imidazolium aprotic ILs react with copper, with different results as a function of the anion.

Zhang et al. [20] reported the corrosion inhibition effect of 1-butyl-3-methylimidazolium chlorides (BMIC) and 1-butyl-3-methylimidazolium hydrogen sulfate ([BMIM]HSO$_4$) on mild steel in 1 M HCl. As a result, it has been concluded that the inhibiting efficiencies decreased in the order of [BMIM]HSO$_4$ > BMIC. Potentiodynamic polarization studies indicated that addition of both ILs affects both anodic metal dissolution and also cathodic hydrogen evolution reactions. Thus, those ILs could be classified as mixed type inhibitors. Also, they found that the mechanism of ILs corrosion inhibition is following the Langmiur adsorption isotherm with the high value of adsorption equilibrium constant. Since, the absolute values of standard free energy of adsorption (ΔG_{ads}) in presence of the studied inhibitors were calculated to be less than 40 kJ mol^{-1}, it has been expected the inhibitors to be physically adsorbed on the metal surface. The corrosion inhibition properties of three different imidazoline based ILs on aluminium in 1 M HCl and 0.5 M H$_2$SO$_4$ were investigated by Quraishi et al. [33] The weight loss study indicated that the inhibition efficiency increased with increase in the concentration of the inhibitor. Moreover, the mechanism of adsorption followed the Langmuir isotherm and behaved as mixed type inhibitors. The most extensively studied IL is based upon the imidazolium cation [31,34,35]. It was observed that the high inhibition efficiency of such inhibitors depends on the specific interaction between the functional groups of IL and the metal surface, due to the presence of the –C=N– group and electronegative nitrogen in the structure of the imidazolium coating [36]. Our indepth review of recent literatures shows that it is important to understand and establish the relation between ILs molecular structure, the counterion, the length of substituted alkyl chains and the functional groups adsorbed on the metallic surface and corrosion inhibition. In next section, the effect of ILs' structure on the inhibition performance has been presented with literature examples.

2.1. Effect of IL Structure on Corrosion Inhibition

2.1.1. Cation Effect

Among many kinds of ionic liquid, imidazolinium and pyridinium cations based ILs have been investigated intensively. Shi et al. [37] synthesized a series of new imidazolium ionic liquids and investigated the relationship between the alkyl connecting with N(3) of imidazolium ring and corrosion inhibition performance in acidic solution. The inhibition efficiency was found to increase with increasing the carbon chain length of the alkyl connecting with N(3) of imidazolium ring. In another study, the corrosion inhibition behaviour of three synthesized imidazolium ionic liquids with similar chemical structure (namely 1-butyl-3-methylimidazolium chlorides, 1-hexyl-3-methylimidazolium chlorides and 1-octyl-3-methylimidazolium chlorides) on aluminum in hydrochloric acid has been investigated [36]. It has been reported that corrosion of aluminum in aqueous solution depends on the concentration of anions in solution. A general mechanism for the dissolution of aluminum as reported in the literature is [38]:

$$Al(s) + H_2O \rightleftharpoons AlOH_{ads} + H^+ + e^- \tag{1}$$

$$AlOH_{ads} + 5H_2O + H^+ \rightleftharpoons Al^{3+} + 6H_2O + 2e^- \tag{2}$$

$$Al^{3+} + H_2O \rightleftharpoons [AlOH]^{2+} + H^+ \tag{3}$$

In the presence of chloride ions the reaction corresponds to:

$$[AlOH]^{2+} + Cl^- \rightarrow [AlOHCl]^+ \tag{4}$$

It is well known that imidazolium group as well as nitrogen atom in heteroaromatic ring of imidazolium compounds can be protonated in acidic solutions. Thus, the interaction of the protonated imidazolium ionic liquid molecules on the aluminum surface competes with the interaction of the ions in the solution. The adsorption of inhibitors on the aluminum surface removes or depletes some of the water molecules originally adsorbed on to the surface, which blocks the formation of $AlOH_{ads}$. Thus, both the oxidation reaction of $AlOH_{ads}$ to Al^{3+} and the complexation reaction between the hydrated cation $[AlOH]^{2+}$ species and chloride ions can be prevented. Moreover, these protonated molecules also compete with the hydrogen ions that can reduce hydrogen evolution. In this case, adsorption would have occurred through polar centers of nitrogen atom in the –C=N– group. Meanwhile, the presence of the electron donating group (Cl) on the imidazolium base IL increases the electron density on the nitrogen of the –C=N– group, resulting in high inhibition efficiency [20]. This particular effect is evidenced more with the increase in chain length of the alkyl connecting with N(3) of imidazolium cationic ring.

In another study, Likhanova et al. [29] reported the inhibition action of imidazolium and pyridinium bromide ILs on mild steel in 1 M H_2SO_4 at room temperature. Since these ILs affected both anodic and cathodic reactions they are classified as mixed type inhibitor. Scheme 2 represents the inhibition mechanism of the interaction between the ionic liquids and the metallic surface. The adsorption of hydronium (H_3O^+) ion and desorption of hydrogen gas (H_2) occurs on the cathodic sites whereas the adsorption and desorption of Br^- and SO_4^{2-} ions occurs on the anodic sites. The protonated imidazolium or pyridinum molecules are electrostatically adsorbed on the cathodic sites in competition with the hydronium ions available to reduce hydrogen evolution [29].

Since cations of ILs are bigger than hydrogen cations, alkyl chain and aromatic ring cover a large part of the metallic surface and lead to the water molecule displacement from surface. Both imidazolium and pyridinum based ILs show a reasonable corrosion inhibition with average corrosion efficiency within 82%–88% at 100 ppm to protect the mild steel corrosion in the aqueous solution of sulphuric acid; their efficiencies are increased with the inhibitor concentration in the range 10–100 ppm. However, due to the larger steric effect of imidazolium ion in comparison to pyridinium, which results in a higher surface coverage area during the chemical adsorption process, imidazolium based IL provides a better inhibition effect than pyridinium.

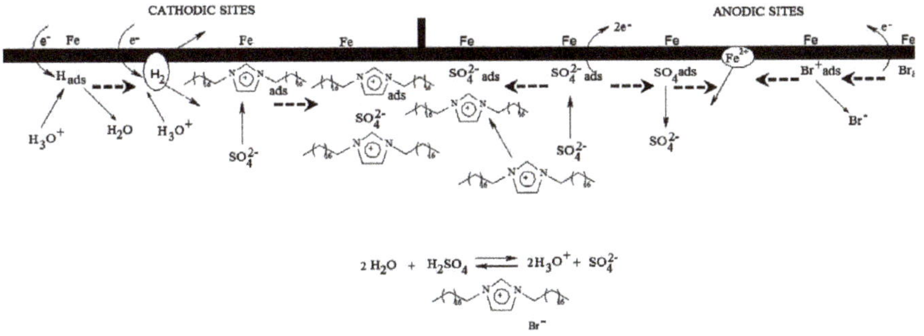

Scheme 2. Corrosion inhibition mechanism of imidazolium and pyridinum molecules on steel surface in 1 M H_2SO_4 (Reprinted with permission from [29]. Copyright 2010 Elsevier).

2.1.2. Anion Effect

For carbon steel and aluminium, the corrosivity of IL media strongly depends on the chemical structure of the cationic moiety and the nature of anion in the IL molecule. Specifically, the corrosion resistance of carbon steel in water-free ILs strongly depends on the IL anion. Depending on the type of IL, carbon steel may undergo severe corrosion in diluted IL media. Anions like tosylate and dimethyl phosphate generally trigger higher corrosivity specifically in water-diluted ILs. In this regard, Uerdingen et al. [39] investigated the behaviour of carbon steel, austenitic stainless steel, nickel-based alloy, copper, brass and aluminium in presence of various diluted ILs with different concentrations under flow conditions at temperatures up to 90 °C. The effect of the chemical structure of the IL cation and the nature of anion on the corrosivity of the metals has been studied. It is observed that diluted ILs (with water) could result in hydrolysis of IL's anions. As a result, they can produce acids (e.g., sulfuric acid, phosphoric acid) and, hence, cause a considerable increase of medium corrosivity, whereas in water-free ILs, most of the metals exhibited a high corrosion resistance.

Ashassi-Sorkhabi et al. [40] studied the effect of IL anion in corrosion inhibitor behaviour of two synthesized imidazolium ionic liquids with chlorides and hydrogen sulphate anion on mild steel in hydrochloric acid. In the structure of the imidazolium bases, the atoms of the imidazolium ring and the –C=N– group can form a big π bond. Then, in addition to the π electron of the imidazolium, bases enter unoccupied orbitals of iron. The π^* orbital can also accept the electrons of d orbitals of iron to form bonds, that produce more than one center of adsorption action. Moreover, the presence of the Cl and S on the imidazolium structure, which are electron donating groups, increases the electron density on the nitrogen of the –C=N– group, resulting in high inhibition efficiency. Therefore, in this study hydrogen sulphate counter-anion showed better inhibiting effect of mild steel in HCl compared with chloride anion.

2.2. Synergistic Corrosion Inhibition Using ILs

As it has been discussed, one of the important advantages of ILs is the ability to select both the anion and cation to have useful properties for a particular application. Due to this feature of ILs and organic salts, they could be used as new organic inhibitors, ideally with synergistic effects. A number of publications have investigated the use of biologically safe anions and cations to produce a salt that could approach the performance of chromates, while being environmentally friendly.

Recently, Somers et al. [41] described such a family of ILs and organic salts that target dual activity by incorporating both anions and cations with proven evidence of effective inhibition. These salts were based on the imidazolinium cation with carboxylate anions. The imidazolinium has a similar structure to imidazolium with a difference in the C4−C5 double bond saturation on the core ring of the imidazolinium. Depending upon the nature of the anion in the salts, these materials were found to have interesting physical properties such as facile ion transport, as well as demonstrating synergistic corrosion inhibition on mild steel. In this study, the influence of pH on the corrosion inhibiting performance of the organic salt for mild steel in chloride environments has been investigated.

It has been shown that this environmentally friendly organic IL remains highly active at pH 2 and 8, which are common environments in which corrosion protection is required. Also at higher pH, the inhibition was controlled by the anion, and the solution showed a high level of protection. Although both the IL's components were mainly ineffective on their own at low pH, the combined salt still had an inhibition efficiency of 72%, indicating a strong synergy between the two ionic species under these conditions. Figure 3 shows an optical microscope image of mild steel samples with corrosion product intact after immersion for 24 h in salt solutions at different pH. Also, at a pH of 8, Figure 3a,b show much less corrosion product, but still show some local attack on the sample in the inhibitor containing solution. At a pH of 2 many bubbles have been observed due to the hydrogen gas evolution, where the sample at pH 2 with inhibitor does not show any bubbles, suggesting a significant reduction in the rate of reaction. The samples immersed in the neutral condition show a similar trend to those at a pH of 8, where the inhibited solution showed much less corrosion product but still with signs of localized attack.

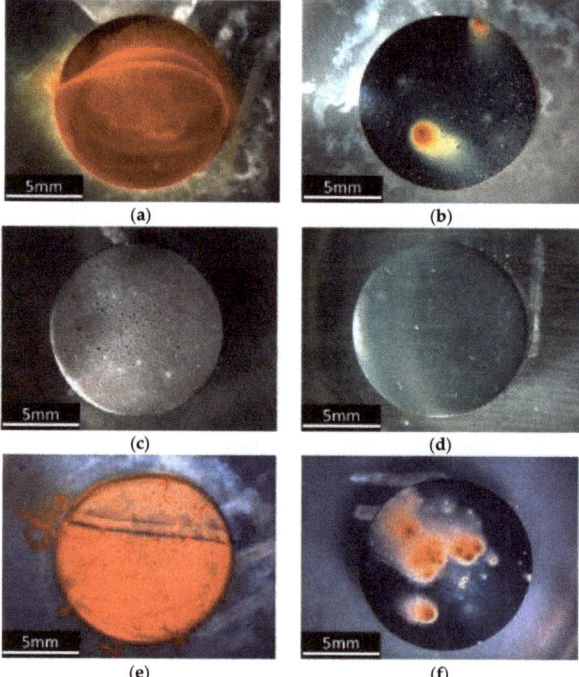

Figure 3. Optical images of the surfaces of mild steel samples after immersion for 24 h in (**a**) NaCl at pH 8; (**b**) NaCl and inhibitor at pH 8; (**c**) NaCl at pH 2; (**d**) NaCl and inhibitor at pH 2; (**e**) NaCl neutral; and (**f**) NaCl and inhibitor in neutral (Reprinted with permission from [41]. Copyright 2016 American Chemical Society).

Such synergy, however, is not always achievable due to some limitation of ILs. One main disadvantages of ILs is unfavourable transport properties of these solvents, which generally present higher viscosity and surface tension than conventional organic solvents [42,43]. Also, once ILs are applied into a coating, they pose problem of miscibility with a coating formulation. For these reasons, polymerized ionic liquids or poly (ionic liquid)s (PILs) are considered more favorable than their monomers in the field of corrosion inhibitor. This is due to their low sensitivity to salts, high shear and thermal stability, high resistivity to strong acid and their efficiencies at lower concentrations. Moreover, such PIL could act as reservoir for IL with controlled release characteristics, such as the microcapsulation of inhibitor which can prevent miscibility issue with other components of coating formulation. Also, they act as controlled release type inhibitor. In the following section detail study of PILs structure and chemistry will be reviewed.

3. Poly Ionic Liquid (PIL) Based Corrosion Inhibitor

A special type of polyelectrolytes which carry an IL species in each of the repeating units are referred to polymerized ionic liquids or poly (ionic liquid)s (PILs), and have been proposed as alternative inhibitor materials. Thus, PILs consist of the cationic or anionic centres on their repeating units in the polymer chain (Figure 4) [29]. Although ILs are in a liquid state near room temperature, PILs are in fact solid in most cases, except a couple of exceptions [44]. Nevertheless, opposite to solid polyelectrolytes, PILs have a reportable glass transition temperature in most cases, being well below usual ionic glasses. The major advantages of using a PIL instead of an IL are the enhanced mechanical stability, improved processability, durability, and spatial controllability over the IL species. The combination of properties of ionic liquids with the flexibility and properties of macromolecular structure results in the unique properties for PILs, which can be used in various applications including solid ionic conductor, powerful dispersant and stabilizer, absorbent, premises for carbon materials, permeable polymers, etc. [45–47]. The initial research of PILs goes back to the 1970s. The major design efforts towards developing novel PILs are focused on vinylimidazolium based PILs with diverse functional substituents due to the positive charge being on an aromatic ring and adjacent to the vinyl groups [48–51]. Intensive studies on PILs significantly expanded the research scope of PILs. New structures, properties and applications have been spotted, which generate several valuable branches for researchers. Meanwhile, there are numbers of reviews, which discussed the synthesis of some PILs and introduced the application of PILs, especially imidazolium based PILs in the field of polymer science [45,52].

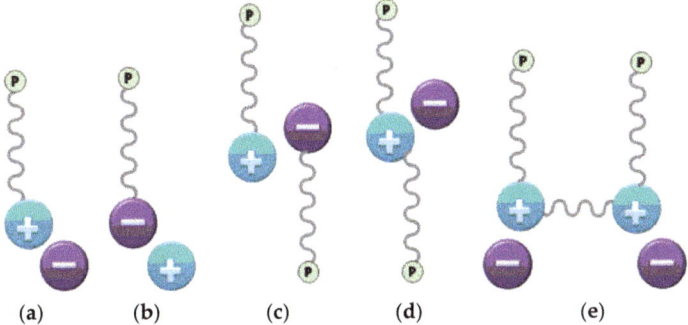

Figure 4. Basic polymerizable IL monomers. "p" represented of a polymerizing unit in an IL monomer. One polymerizable unit located on the cation (**a**) or anion (**b**). Two polymerizable units separated in the cation and anion (**c**), or located on the cations (**d,e**).

It should be noted that despite the very interesting properties of PILs and their wide range of applications, these eco-friendly compounds have received little interest as corrosion inhibitors. It has been reported that some ILs based on imidazolium, pyridinium and pyridazinium exhibited corrosion inhibition properties for the corrosion of various metals [21], however; there is very limited investigation for application of PILs as a corrosion inhibitors. Here, a few recent work in this area will be discussed. Olivares-Xometl and co-workers reported the poly(ionic liquid)s (PILs), derived from imidazole with different alkylic chain lengths for corrosion inhibition of aluminum alloy in diluted sulfuric acid [53]. Figure 5 shows the likely mechanism of PILs' interaction with both the metallic surface and the aggressive environment. The interaction among the hydrophobic parts of the PILs molecules could support the protective action. However, it is more likely that the main chains of the polymer form an obstacle, which may have a supportive action on inhibition, as they hinder the passage of water and aggressive ions, in agreement with their hydrophobic nature. When the alkyl side chain is composed of 12 carbons, more effective steric hindrance prevalent, as it can interact with the other lateral alkylic groups to slow molecular diffusion.

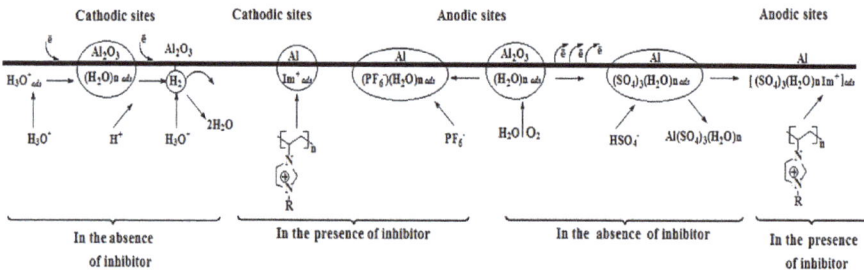

Figure 5. Schematic representation of corrosion and inhibition of aluminum alloy before and after PIL addition in diluted sulfuric acid [53].

However, in this study PILs displayed a short protection range for the alloy. Thus, these PILs are not suitable to be applied in acidic media, as they are not easily adsorbed due to ionic competition, which leads to the formation of a non-uniform corrosion inhibitor film on the aluminum alloy surface. In another study, by Ayman et al. [54,55] PIL based on 2-acrylamido-2 methylpropane sulfonic acid, showed an excellent corrosion inhibition performance for steel in 1 M HCl medium. The adsorption of IL on steel surface blocked the active centers, which lowered the corrosion rate of steel. It has been reported that introducing oxyethylene ammonium counter ion into the ionic liquid polymer system promotes the wetting characteristics to form anticorrosion protective layer at the solid surfaces [54]. Also, it has been indicated that PIL in this system behaved as a mixed type inhibitor and acted via adsorption on steel surface by hindering and retarding the active centers from the corrosion reaction.

3.1. PIL Structure Diversity

There have been persistent efforts devoted to the preparation of PILs in various forms and dimension scales like spherical micro-/nanoparticles, micro-/nanogels, vesicles, nanoworms, etc. [56]. Therefore PILs have a diverse chemical structure reservoir. The most recent forms of PILs, which have been mostly used in different applications, are PIL colloidal particles and PIL gels. An insight into these types of PILs and their characteristic features is presented in the following section.

3.1.1. PIL Colloidal Particles

Since colloidal systems have a close connection with nature and human life, PIL colloids are a new platform to investigate the unique properties of ILs with the rather small dimension and the superior dispersity of colloidal particles. Moreover, they setup a powerful platform for a variety of studies on PIL functions and applications. A colloidal system is defined as a state of subdivision dispersed in a medium with at least one dimension between approximately 1 nm and 1 μm [57]. Recently, nanostructured PILs, and especially colloidal nanoparticles, have received significant interest as functional polymer nanoparticles. This is due to the fact that small particle sizes increase surface effects on the interfacial interaction and mass/energy transport. Moreover, a small particle size and the charged character of PILs improve the colloidal stability in aqueous as well as non-aqueous dispersions. So far, a few synthetic routes have been developed to prepare PIL nanoparticles including suspension polymerization [58], water-in-oil concentrated emulsion polymerization [59], and precipitation polymerization in water without stabilizers using ionic liquid monomers with long alkyl chains has been recently reported [60,61]. Among PIL nanoparticles a significant focus in this field is on the nanostructured imidazolium-type PILs [47,62]. One of the examples of this type of applications is the work by Yang et al. [63], which reported the preparation of the crosslinked poly(1-butyl-3-vinylimidazolium bromide) microspheres with the diameter of about 200 nm synthesized via miniemulsion polymerization for application as metal scavenging and catalysis.

In another work by Zhou et al. [64], a thermosensitive type ionic microgels was obtained via the surfactant-free emulsion copolymerization of 1-vinylimidazole and 4-vinylpyridine with thermosensitive monomer N-isopropylacrylamide. The obtained microgels were spherical in shape with narrow size distribution and exhibited thermosensitive behaviour with unique features of PILs in aqueous solution (Figure 6).

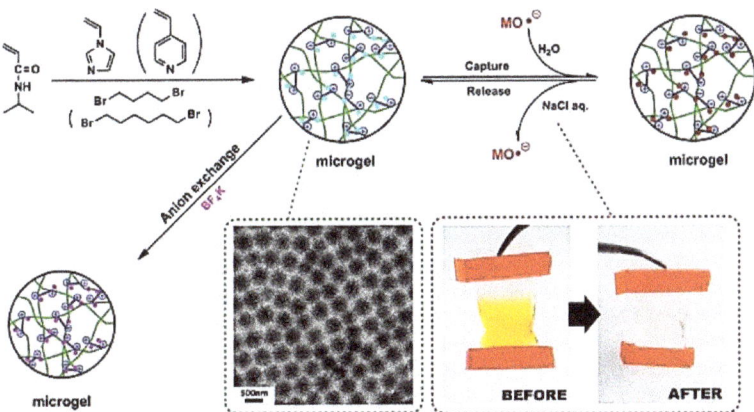

Figure 6. Schematic summary of synthesis of thermosensitive ionic microgels via quaternized crosslinking reaction and their properties (Reprinted with permission from [64]. Copyright 2014 American Chemical Society).

One of the most common methods used for the fabrication of polymer nanoparticles with the droplet sizes typically in the range of 20–200 nm is miniemulsion. The use of water as the dispersion medium is one of the main advantages of this system, which makes it environmentally friendly and also allows excellent heat dissipation during the polymerization process [65]. On the other hand, very recently, "click" polyaddition reactions in miniemulsions, specifically thiol-mediated chemistry (i.e., thiol-ene/yne, thiol-Michael); have attracted attention as one of the facile methods for synthesis of polymer nanoparticles, and nanocapsules dispersion in heterophase media with high efficiency [66,67]. For example, for the first time Jasinski and co-workers reported the preparation of poly (thioether ester) latex nanoparticles using miniemulsion thiol-ene photopolymerization. Their synthesis was performed in water, at ambient temperature, and without the use of any organic solvent. The resultant linear poly (thioether ester) particles had an average diameter of 130 nm [68].

In another recent work, sub-100 nm crosslinked polythioether nanoparticles were synthesized via thiol-ene photopolymerization in miniemulsion using high-energy homogenization [65]. Our group recently reported the facile preparation of cross-linked PILs based nanoparticles via thiol-ene photo-polymerization in miniemulsion [69]. In this study, the PIL nanoparticles exhibited improved corrosion inhibition properties to the sol-gel coating due to the interaction between the –C=N– group and electronegative nitrogen in the PIL with the metal surface.

3.1.2. PIL Gel

As described above, PIL gels are one form of PILs which recently have been used in a number of applications such as electrolytes for batteries and supercapacitors [70], drug delivery [71], agriculture, and biomedical fields [72]. Indeed, they are showing a multitude of characteristics that make them very versatile materials with tuneable properties. Gordon and co-workers pioneered the synthesis of large PIL beads in the micron meter size scale through direct polymerizations of 1-butyl-3-vinylimidazoium TFSI in presence of a 1,8-di(vinylimidazolium)-octane TFSI as a crosslinker. Furthermore, the resulting gel-type beads were swelled in acetone, and loaded with palladium nanoparticles to catalyze C–C coupling reactions [58]. The authors suggested the application of such gel beads in catalysis, separation technology, and ion-exchange resins. Xiong and co-workers [73] reported a facile one-step synthetic strategy for the preparation of cross-linked polymeric nanogels by the conventional radical copolymerization of a phosphonium-based IL for use as catalysts. Recently, Rahman et al. used a microfluidic method to fabricate monodisperse spherical PIL microgel beads [74]. The authors showed the anion exchange can enable fine-tuning of size and swellability of these beads. By incorporating diverse anions, they were able to impart a multitude of functionalities to these beads, ranging from redox capabilities, controlled release of payload, magnetization, toxic metal removal and robust, reversible pH sensing. These chemically switchable stimulus-responsive PIL beads have potential applications in portable and preparative chemical analysis, separations and spatially addressed sensing (Figure 7) and also have potential for use as cargo for corrosion inhibitors or slow release inhibitor.

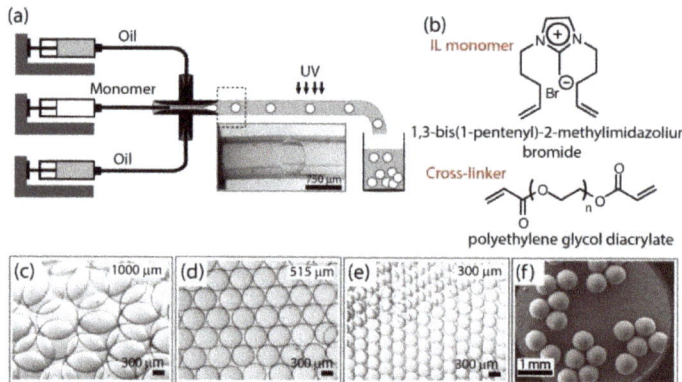

Figure 7. Schematics illustrating microfluidic method to generate switchable stimulus-responsive PIL microgels. (Reprinted with permission from [74]. Copyright 2013 American Chemical Society). (**a**) Stereomicroscope image of a prepolymer droplet flowing in the transparent capillary tube; (**b**) Chemical structures of IL monomer and cross-linker; (**c**–**e**) Stereomicroscope images of PIL microgels showing their monodispersity and transparency and (**f**) FESEM image of synthesized PIL[Br].

With the aim of using such PIL as corrosion inhibitors, we reported the novel and facile fabrication of multifunctional PIL gel beads using vinyl imidazolium based ionic liquid through click-type reactions (Figure 8) [75]. A detailed study into the effect of reactant ratios is examined.

The gel formation is confirmed through fourier transform infrared spectroscopy (FTIR), thermal analysis, and kinetic studies. These PIL gels exhibited multiple characteristics including (1) self-healing characteristics due to their rubbery nature, (2) the ability to uptake active molecules which acts as corrosion inhibitors, and (3) pH sensing through the incorporation of indicator molecules. These functionalities demonstrate the potential of PIL gel family as multifunctional autonomous platform material for the control, detection and inhibition of corrosion.

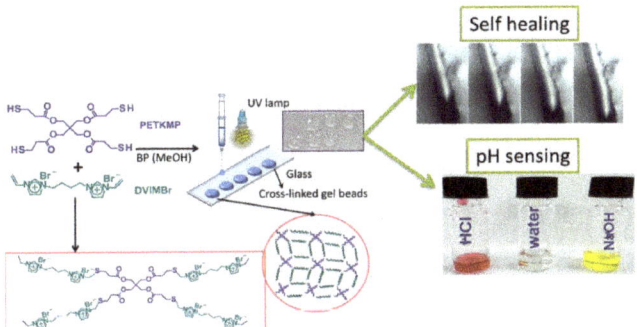

Figure 8. The facile fabrication of multifunctional PIL gel beads using vinyl imidazolium based ionic liquid through click-type reactions. (Reprinted with permission from [75]. Copyright 2015 American Chemical Society).

4. Graphene as Green Corrosion Inhibitor in Anticorrosion Coating

Graphene is a nanofiller with one-atom-thick planar sheet of two-dimensional carbon with sp^2 bonded carbon atoms that are densely packed in a honeycomb crystal lattice or an unrolled single-walled carbon nanotube [76]. Different approaches for preparing graphene sheets have been investigated like graphite exfoliation, including mechanical cleavage of graphite, chemical exfoliation of graphite, thermal-induced exfoliation, and direct synthesis, such as epitaxial growth, and bottom-up organic synthesis. Prasai et al. [77] studied the corrosion inhibition effect of copper and nickel by either growing graphene on these metals by chemical vapor deposition (CVD) method, or by mechanically transferring multilayer graphene onto them (Figure 9). Graphene grown by chemical vapour deposition (CVD) technique has shown superior anticorrosion coating but it is also demonstrated that these coating cannot be used over a long-term duration. It has been reported that transferring multiple graphene layers onto the metal surfaces will increase the degree of protection with building thicker and more robust films.

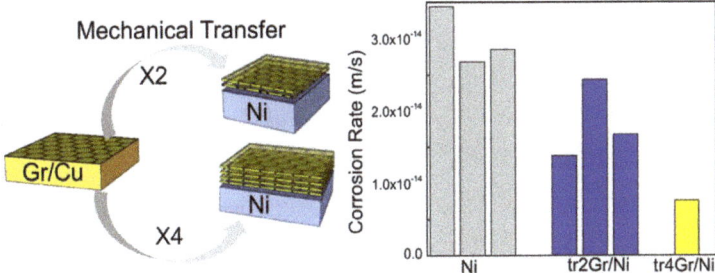

Figure 9. Schematic demonstration of thin layers of graphene as a protective coating that inhibits corrosion of underlying metals. (Reprinted with permission from [77]. Copyright 2012 American Chemical Society).

High thermal conductivity, better gas barrier, extraordinary electronic transport properties, superior mechanical stiffness combined with a wide set of other unusual properties of graphene-based composites made them promising and cheaper alternative to carbon nanotubes-based composites [78–81]. Graphene and graphene derivatives (e.g., graphene/graphite oxide, functionalized graphene, etc.) could be used in various applications such as hydrogen storage [82], sensors [83], transparent conductive films [84], batteries [85,86], super capacitors [87], solar cells [88] and nanocomposites coatings [89–91]. Due to the high surface area of graphene sheet (2630 m^2/g), improvement of mechanical, thermal, and electrical properties of composite graphene based coating could be achieved with very low loading [92]. Chang et al. [93] applied polyaniline/graphene composites (PAGCs) for corrosion inhibition of steel. The composites display outstanding barrier properties against O_2 and H_2O. Figure 10 depicts the corrosion inhibition behaviour of bare steel and PAGCs coated steel with different amount of graphene loading in a corrosive medium (3.5 wt % aqueous NaCl electrolyte) under potentiodynamic polarization conditions. As it can be observed, as the PAGCs loading was increased further, the corrosion inhibition ability was enhanced evidenced by the highest E_{corr} and lowest I_{corr} values (which corresponds to a lower corrosion rate). In fact, using graphene in coating matrix could increase the length of the diffusion pathways for reactive gases such as oxygen and water vapour in polymer coatings and lead to a remarkable improvement of the corrosion inhibition of metallic substrate compared to normal polymer coating.

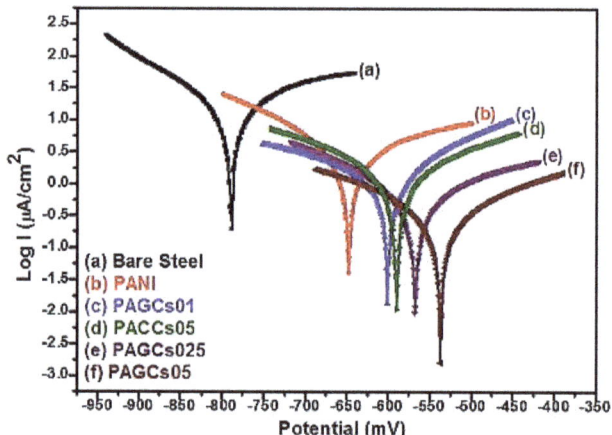

Figure 10. Tafel plots for (a) bare steel; (b,c,d,e,f) PANI-coated with different amount of graphene loaded. Electrodes measured in 3.5 wt % NaCl aqueous solution (Reprinted with permission from [93]. Copyright 2012 Elsevier).

Wang et al. [79] suggested that incorporation of graphene sheets into the epoxy polymer composite improved thermal conductivity and reduced coefficients of thermal expansion (CTEs). Their results also indicate that due to the high thermal-stability of graphene, they can be used in microelectronics coating applications. Since it is easy to obtain the graphene precursor, graphite, as it is naturally abundant, and the functionalized graphene can serve as a conductive nanofiller for other polymers (such as epoxy, polyimide, polyurethane, etc.), polymer/graphene based composite coatings will emerge as a new area of corrosion inhibition technology.

Stronger interface have been achieved using graphene platelets (GP) comprising one or more layers of a graphene plane. Yasmin et al. [94] have developed epoxy/graphite nanocomposites by mixing epoxy with graphite in solvent; 4 wt % graphite increases Young's modulus by 10% and glass transition temperature (T_g) marginally from 143 to 145 °C. Better results have been obtained using sonication and shear mixing, 1 wt % GP increasing modulus 15%, but leads to a reduction of tensile strength. The mechanical properties of epoxy/GP nanocomposites have been investigated by Koratkar et al. [95] showing improvement in epoxy fracture toughness from 0.97 to 1.48 MPa m$^{1/2}$ at 0.1 wt % filler fraction. Therefore, it could be used as toughening agent for coating.

Despite the tendency of graphene nano-sheets to re-aggregate and stack due to their high surface area and strong van der Waals force has limited their applications in polymer nanocomposites. However, several studies have focused on improving the dispersion and interface interaction of graphene in a polymeric matrix using functionalised graphene. Novel method for functionalization of GP has been presented by Miller [96] using a coupling agent to form covalent bonding between fillers and soft matrix (0.78 GPa Young's modulus), resulting in 50% modulus improvement at 1 wt % filler fraction. Chiang and Hsu [97] have improved the fire resistance of epoxy/GP nanocomposite following a similar method. Martin-Gallego et al. [98] studied the effect of functionalized graphene sheets (FGS)/epoxy coatings which are prepared using cationic photopolymerization on mechanical properties of coating. Their results indicate increased stiffness and T_g values of the cured epoxy network with better storage modulus properties in higher temperature.

Jeong and co-workers [99] investigated the effect of graphene content on structures and electrical properties of graphene/epoxy composite films which are prepared by solution casting and following thermal curing of diglycidyl ether of bisphenol-A with an amine-functionalized agent mixed on a polyimide film. The graphene/epoxy composite films can be utilized as high performance electric heating elements in various applications. They found that the graphene content as well as the applied voltage are two key elements in controlling the maximum temperatures of the composite films. Bao et al. [100] enhanced the mechanical, electrical and thermal properties of the epoxy nanocomposites utilizing functionalized graphene oxide. In situ thermal polymerization has been used to functionalize graphene oxide (FGO) via surface modification by hexachlorocyclotriphosphazene and glycidol. Strong interfacial interaction between FGO and epoxy matrix improved the thermal stability, storage modulus and hardness in a polymeric matrix.

Thus, graphene as an anti-corrosive agent is very attractive as it may protect metals by keeping their intrinsic properties, which cannot be achieved using three dimensional protective paints, oxides or polymers. In the field of using graphene as corrosion protective material the biggest hurdle is that the graphene sheets synthesised using current methods still contain too many defects. So the main challenge in near future is to improve the quality of sheet produced, the poor quality of sheet drastically reduces the performance as an anti-corrosive material. The keys factors affecting the quality of sheets are defects or abnormalities in graphene sheets like:

- Missing bonds;
- Pentagonal and hexagonal lattices;
- Lattice distortion;
- Local thickness variations;
- Presence of impurities.

All these factors can represent the centre of damage accumulation also, other than altering the properties of graphene. Local defects can lead to accumulation of oxygen which ruins the chemical properties of sheet. Presently the functionalization of GO via non-covalent and covalent route with organic compounds has become a matter of rigorous research for production of innovative hybrid nano composites with new advanced functions and applications.

Quantum Chemical Methods as Efficient Tools to Study Corrosion Inhibitors

Quantum chemical methods are seen to be very effective in determining the molecular structures as well as explaining the electronic structures and reactivity's of molecules. Computational chemistry is considerably used to assess the efficiency of corrosion inhibitors, this method helps to search compounds of desired property employing computational modelling. Quantum chemical method and molecular modelling techniques help in defining a large number of molecular parameters illustrating the reactivity, shape, and binding properties of complete molecules as well as of individual molecular fragments and substituents. The prominent quantum chemical parameters are atomic charges, molecular orbital energies (E_{HOMO}, E_{LUMO} and ΔE_{gap}), dipole moment, charge distribution. Density functional theory has successfully been applied to explain the importance of structure of corrosion inhibitors and their adsorption efficiency on the metal surfaces, however the properties of corrosion inhibitors like E_{HOMO}, E_{LUMO}, ΔE_{gap}, dipole moment (μ), electronegativity ($\varkappa$), and atomic charge have by far achieved the appropriate correlation with corrosion inhibition efficiency.

5. Emerging Embedment Methods of Corrosion Inhibitors

Corrosion inhibitors can be incorporated into the coating formulation through different ways. One of the most commonly used methods is the direct addition of inhibitors in the primer or topcoat. However, a too high concentration or low solubility of the inhibitors often results in a deterioration of the integrity and physical barrier properties of the matrix of the protection system [101]. In addition, the existing interaction of the inhibiting agents with the protective matrix often leads to significantly reduced stability of the protective layer and the deactivation of the inhibitors. Recently, different new methods for inhibitor incorporation have emerged to prevent the direct interaction of inhibitor with the matrix. One of the most common methods is application of inhibitor loaded coatings. Coatings based on inhibitor loaded containers protect the metal by releasing corrosion inhibitor in response to changes in the coating integrity (cracks) or local environment (pH shift) caused by corrosion attack. These systems have been extensively investigated, because they are potential replacements for the banned chromate-based coatings [102]. Out of these, self-healing coating is an emerging and broad field to replace the chromate for corrosion control and autonomic repair of coatings (self-healing), which is discussed in this section.

5.1. Self-Healing Coating

Coating with self-healing properties is an advanced application of emerging corrosion inhibitors. The concept of self-healing which is initiated in the nineties by Dry [103] and Sottos [104] is the known phenomena seen in the nature and refers to self-repair. Self-healing coatings can be classified into two main classes' namely (1) extrinsic and (2) intrinsic self-healing systems. In extrinsic self-healing systems such as capsule-based and vascular systems, the healing agents are added as a separate phase into the matrix, while intrinsic systems such as ionomers, hydrogen-bonded systems, etc., are those which are free from healing agent and do not require any external energy to trigger the response [105,106]. They can repair the mechanical damage spontaneously due to the architecture of the molecules themselves and avoiding rupture and corrosion of underlying substrate. Extrinsic technique possesses several advantages over intrinsic, which will be discussed in this section.

In contrast to conventional anticorrosion coating, emerging corrosion inhibitor embedded in self-healing coating can act in response to corrosion attack, decrease the corrosion rate thereby enabling less maintenance and durability of the coating. For achieving this goal, the coatings have to provide release of the active and repairing material rapidly after integrity changes in coating. The main idea is to load active agents (e.g., corrosion inhibitors) into nanocontainers surrounded by a shell which controlled the permeability and then to introduce them into the coating matrix. Consequently, nanocontainers are keeping corrosion inhibitor in a "trapped" state and distributing uniformly in the passive matrix. Thus, the undesirable interaction between the corrosion inhibitor (active material) and the passive matrix which leads to spontaneous leakage could be prevented. When the local environment undergoes changes or if the active surface is affected by the outer impact, the nanocontainers respond to that signal and release encapsulated inhibitor. Various methods to add the self-healing properties to coatings have been investigated including encapsulation, reversible chemistry, microvascular networks, nanoparticle phase separation, polyionomers, hollow fibres, and monomer phase separation [107]. Microvascular is a strategy in which material with interconnected series of network channels has been designed. In this approach, circulatory system continuously transports the necessary chemicals and building blocks of healing to the site of damage. Therefore, coating on a substrate containing a micro channel network is healed. This is the most biomimetic approach and it is difficult to achieve practically and at large scales in synthetic materials. Nanotubes are another approach that may be able to deliver larger amounts of liquid healing agent to the crack plane. Halloysites (aluminosilicate nanotubes) which are one of the most abundant natural nanotubes have recently been applied as containers in the automotive and maritime industries for corrosion protection. They have been developed as an entrapment system for loading, storage, and controlled release of corrosion inhibitor in coatings [108]. The drawbacks of using nanotubes lie in its poorly defined composition and its narrow particle size [107]. So far the most successful approach in self-healing the polymeric component of organic coatings is microencapsulation. This approach has significant advantages including protection of reactive materials (inhibitors) from corroding environment, controlled evaporation of inhibitors, safe handling of toxic inhibitors and controlled release of the inhibitors for delayed release or long acting release [109].

5.1.1. Encapsulated Type Self-Healing

Corrosion inhibitors can be easily embedded into the capsules through variety of techniques. These techniques can be merged into two main categories; physical and chemical (Figure 11) [1]. There are different chemical approaches for synthesizing the microcapsules such as interfacial polymerization [110], coacervation, in situ polymerization [111,112], extrusion, and sol-gel methods. The fastest and most convenient method among them is in situ polymerization. In this approach, microcapsules containing healing agent (inhibitor) in a trapped state disperses uniformly in the matrix containing a catalyst capable of polymerizing the healing agent and fracture upon loading of the coating, releasing the low viscosity self-healing reagents to the damaged area for curing and filling of the micro cracks (Figure 11) [113,114].

Brown et al. [112] are known as a pioneer of the micro-/nanocapsules synthesis with their achievement of micro capsulation of dicyclopentadiene (DCPD) as a healing agent with urea–formaldehyde (UF) shell using in situ polymerization. Most commonly used healing agents as a core are dicyclopentadiene (DCPD), epoxy, linseed oil, tung oil, o-dichlorobenzene and dimethyl siloxane. Shell materials are mainly limited to poly(urea–formaldehyde) (PUF) and poly(melamine–formaldehyde) (PMF) or melamine modified poly(urea–formaldehyde) (MUF) [114]. Table 1 summarises recent work that has been carried out in microcapsulation based self-healing system for coating applications.

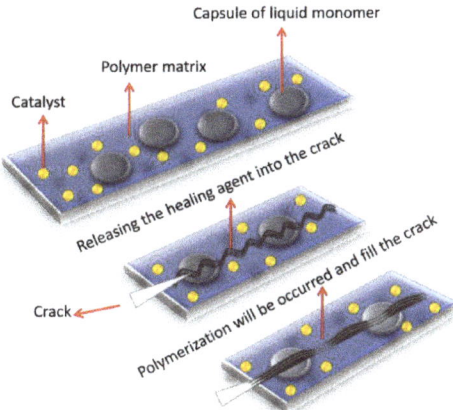

Figure 11. Encapsulated type self-healing through in situ polymerization technique.

Table 2 is the summary of characteristics which will be required for designing microencapsulation-based self-healing polymeric materials [115]. It is crucial to consider four steps process of achieving healing ability to obtain better functionality: storage, release, transport, and rebonding. Each of these steps depends significantly on the chemistry and properties of the healing agent system [107]. Table 2 reveals the importance to develop capsules with good compatibility with the coating matrix and considering the possibility to encapsulate and upkeep active material, and control of the permeability properties of the shell through external stimuli. Shell permeability could be changed reversibly or irreversibly by various stimuli: variation of the pH, ionic strength, temperature, ultrasonic treatment, alternating magnetic field and electromagnetic irradiation.

As a result, different responses can then be observed, such as tuneable permeability or more drastic ones like total rupture of the container shell. Also as it has been shown in this table, size of the capsules is another important parameter which should be less than 300–400 nm; capsules of larger size can reduce the protective performance of the coating [107].

Table 1. Summary of most recent work in microcapsule based self-healing.

Microcapsule Components (Core and Shell)	Chemistry	Specific Feature	Size of Capsule	Ref.
Shell: polysulphone Core: 1-hexyl-3-methylimidazolium bis(trifluoromethylsulphonyl) imide [HMIM][NTf2] ionic liquid	Solvent evaporation.	Chemically stable within the high-temperature curing conditions necessary for the coating system (up to approximately 380 °C).	Below 10 μm	Magalhães et al. [116]
Shell: epoxy–amine(ethylenediamine (EDA)) Core: epoxy	Interfacial polymerization	improved compatibility and adhesion with the coating matrix especially if the coating is alkaline	100 μm	Liu et al. [110]
Shell: poly(urea–formaldehyde) Core: 1H,1H,2H,2H-perfluorooctyl triethoxysilane (POTS)	In situ polymerization	good corrosion protection ability to steel; self-healing behaviour was realised under ambient condition without any manual intervention	100 μm	Huang et al. [111]
Shell: poly(urea-formaldehyde) Core: octyldimethylsilyloleate	In situ polymerization	great potential of the silyl esters as healing agents and good results in corrosion protection	50 and 100 μm	García et al. [117]
Shell: ethylene glycol dimethacrylate (EGDM) Core: ionic liquid, 1-hexyl-3-methylimidazolium bis(trifluoromethane sulfonyl)amide	Self-assembling of phase separated polymer (SaPSeP method)	ionic conductivity; good results in corrosion protection	Multi hollow structure	Okubo et al. [118,119]

Table 2. Characteristics required for designing microencapsulation-based self-healing polymeric materials.

Component	Characteristics
Corrosion inhibitor	Stability and shelf-life Deliverability Reactivity Shrinkage Physical and mechanical properties Thermal stability
Microcapsule shell wall	Chemical compatibility Mechanical properties Dispersion Thermal stability
Catalyst, curing agent, or reaction initiator	Solubility Chemical compatibility Reactivity Dispersion Thermal stability

5.1.2. Effective Parameters and Challenges of Microcapsule Embedment for Corrosion Inhibition

It is evident that application of self-healing coatings will be the most common and cost effective method of improving the corrosion protection. However, for the excellent fabrication of self-healing coatings several parameters must be considered such as inhibitor material, microcapsule diameter (size), microcapsule core and shell, microcapsules dispersion, presence of catalyst, coating application, coating thickness and coating matrix. Therefore, there is a growing need for investigation of effective parameters in microcapsule formation which is under intense study by various researchers. For example, Nesterova et al. [120] found that an increase in stirring rate, stirrer geometry, correct choice of temperature, and a high stabilizer concentration all can affect the microcapsule size. In capsules with irregular shape mechanical stability will be compromised and capsules will be unacceptable for a coating use. Another interesting and effective parameter which should be considered for obtaining self-healing property is the position of capsules in the coating matrix. Therefore, Kumar et al. [121,122] studied two methods of applying microcapsules in the primer layer: mixing to the primer before applying and sandwiching the microcapsules in the primer during application. Experimental results suggested that the microcapsules should be mixed into the paint formulations at the time of application.

In the other work by Cho et al. [113], two self-healing systems based on siloxane materials have been studied. First system consists of phase-separated polydimethylsiloxane (PDMS) healing agents and microencapsulated catalyst. The limitation of this system is the possible reaction between PDMS healing agent and coating matrix. In order to overcome this drawback, in the second system both catalyst and PDMS healing agent are encapsulated within urea-formaldehyde (UF) microcapsules. Their dual-capsule PDMS healing system showed no evidence of corrosion in the damaged area even after a long time exposure to the corrosive species. Although self-healing or autonomously healing micro cracks is a promising approach for extending the life of coating, still there are significant of unsolved challenges for optimization of the autonomous microcapsule system which is suitable for multiple healing actions.

6. Evaluation of Corrosion Inhibitors Using Advanced Characterization Techniques

Electrochemical methods are most commonly used techniques for the evaluation of the efficiency of corrosion inhibitors. The advantages of electrochemical methods are short measurement time and mechanistic information that they provide which help not only in the design of corrosion protection strategies but also in the design of new inhibitors. Although several electrochemical techniques may be used to study the performance of corrosion inhibitors, potentiodynamic polarization method and electrochemical impedance spectroscopy (EIS) can provide significant useful information, which makes them the most useful method for such study and the number of reports used this method for study of corrosion inhibition performance is limited. As an example, Otmacic Curkovic et al. studied the mechanism of the protective action of three imidazole-based (4-methyl-1-(p-tolyl)-imidazole, 4-methyl-1-(o-tolyl)-imidazole, and 4-methyl-1-phenyl imidazole) corrosion inhibitors on copper in 3% NaCl, using quartz crystal microbalance measurements [123]. This study confirmed that even slight changes in the molecular structure induce a significant effect on the inhibiting properties. Both tolyl-substituted 4-methyl imidazoles rapidly adsorbed onto the copper surface and decreased the copper corrosion rate while the phenyl-substituted 4-methyl imidazole slowly formed a protective 3D layer. On the other hand, the inhibiting effect of o-tolyl-substituted compound did not improve with time, while the inhibiting efficiency of the phenyl-substituted inhibitor increased with immersion time. Figure 12 presents the mass and the corrosion potential changes with respect to immersion period measured by QCM-D in the presence of (a) 4-methyl-1-(o-tolyl)-imidazole and (b) 4-methyl-1-phenyl imidazole. The use of better inhibitor (phenyl based imidazole) shows the increase of the mass of the electrode which is due to the formation of protective layer on the copper surface. However, very little work has been performed on GO/IL systems.

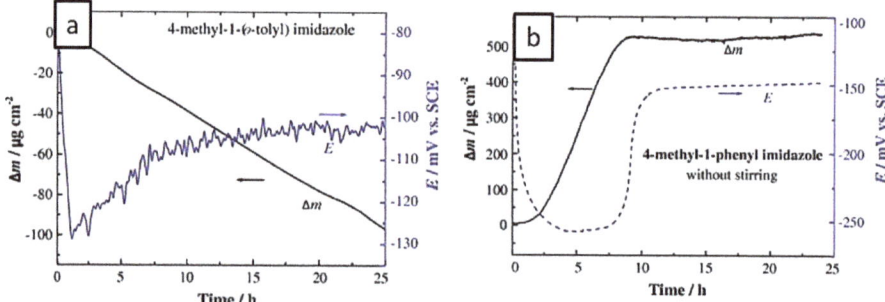

Figure 12. Mass and potential change of copper electrode in 3% NaCl with the addition of (**a**) 4-methyl-1-(p-tolyl)-imidazole and (**b**) 4-methyl-1-phenyl imidazole (Reprinted with permission from [123]. Copyright 2009 Elsevier).

On the other hand, the measurement of electrochemical reactions at the interface becomes a matter of particular interest for the prediction of the service life of coating. To achieve such information, new techniques that perform local measurements, such as scanning kelvin probe SKP are increasingly applied [124,125]. This method is a non-destructive, non-contact mode technique based on a vibrating capacitor to measure the surface work function (WF) distribution on the coating surface. Surface work function is an extremely sensitive indicator of the surface condition and can be used to track changes in the surface such as surface contamination and corrosive adhesion of polymers on metal substrates [126].

Among many studies, Choudhury et al. [127] recently presented the sol-gel derived hybrid coatings containing three different compositions of methacrylate-phosphosilicate on mild steel substrates where SKP microscopy was used to evaluate the adhesion and corrosion protection properties. Following equation can be used to correlate the absolute WF measured using SKP (V_{KP}) to the corrosion potential (E_{corr}) [128]:

$$V_{KP} = E_{corr} + \text{const.} \qquad (5)$$

Phosphorus containing methacrylate hybrids were synthesized from 2-(methacryloyloxy)ethyl phosphate (EGMP) and 3-[(methacryloyloxy)-propyl] trimethoxysilane (MEMO) via dual-cure process involving sol-gel reaction and addition polymerization. Similar experimental procedures were used to synthesize hybrids at other composition namely M:E–3:7 [129]. Figure 13 illustrates the SKP maps of the gold-coated aluminum (reference), bare metal substrate (MS), MEMO, EGMP, M:E–1:1 and M:E–3:7 coated samples. The average WF values of the samples shift positively towards noble potential in the order of bare MS < EGMP < MEMO < M:E–1:1 < M:E–3:7.

The deviation in WF values can be correlated to the interfacial interaction between the coating and the substrate. The SKP measurements showed the presence of strong interfacial interaction, which is attributed to the interaction of phosphate group with the metallic substrate [130].

An understanding of the correlation between the structure and observed corrosion inhibition properties, such as mechanism of adsorption, is essential for designing corrosion inhibitors with enhanced properties. Adsorption of an inhibitor on a metal surface depends on various parameters such as the nature and the surface charge of the metal, the inhibitor's chemical structure etc. One of the most common scattering techniques used for corrosion and corrosion inhibition study is surface-enhanced Raman scattering [131]. However, this technique studies the film that forms on the metal and is a surface analysis technique.

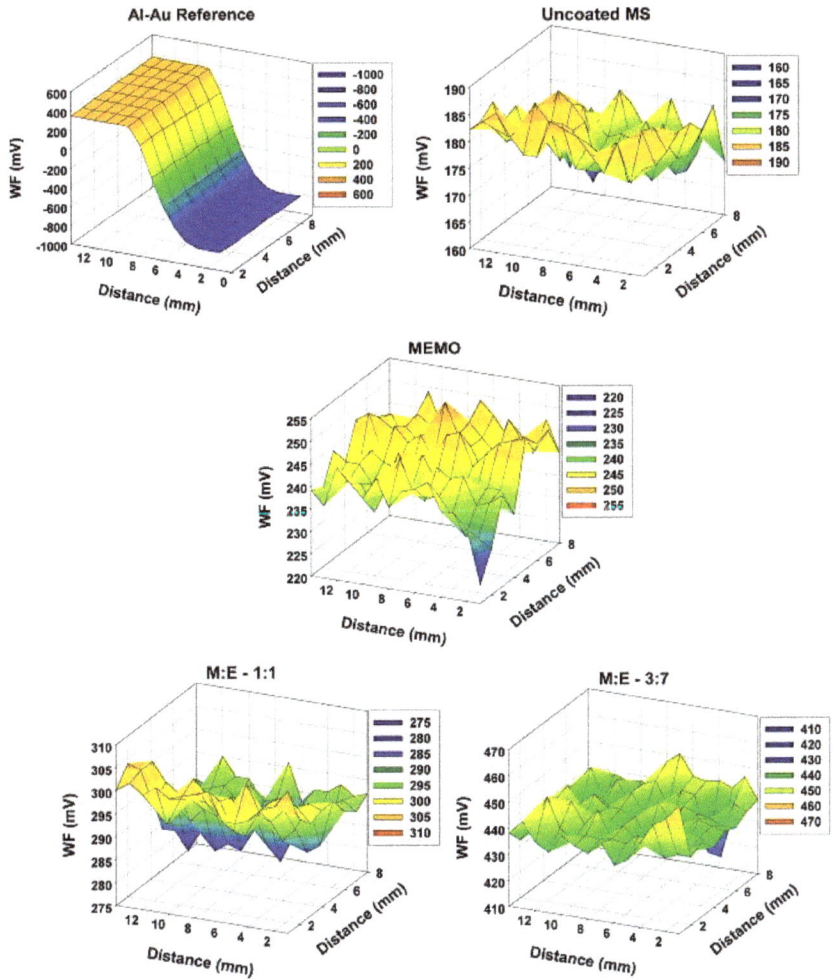

Figure 13. SKP maps of pristine gold on aluminum reference, bare MS, MEMO, M:E–1:1 and M:E–3:7 coated samples, respectively. (Reprinted with permission from [127]. Copyright 2010 Elsevier).

In the field of organic corrosion inhibitors, more attention is paid to the mechanism of adsorption and also to the relationship between inhibitor structures and their adsorption properties. Neutron scattering techniques including small angle neutron scattering (SANS) and ultra-small angle neutron scattering (USANS) are valuable techniques to study the structure of corrosion inhibitors in different forms including gels and nanoparticles. SANS and USANS are ideal and can be used to provide information relating to the crosslink porosity of structure, which is directly related to the mechanical properties of inhibitors. A literature survey of neutron scattering demonstrates that this technique has been employed to study the microstructure of a range of different types of polymers, predominantly from synthetic polymers. Furthermore, polymers studied by this method are classified into polymer blends, block copolymers and polymer gels [132].

Shibayama et al. [133] studied the structure of tetra-arm polyethylene glycol (PEG) gels by SANS. It has been investigated that there is no inhomogeneities appeared even by swelling. However, a steep upturn was observed in SANS curves, indicating the presence of PEG chain clusters or defects where these inhomogeneities disappear in swelled sample. Furthermore, Bhatia and co-workers designed a unique form of chemically cross-linked PEG gels to minimize defects in the network [134]. SANS was utilized to investigate the network structures of gels in two different solvents: D_2O and d-DMF. SANS results show the resulting network structure is dependent on PEG length, transitioning from a more homogeneous network structure at high molecular weight PEG to a two phase structure at the lowest molecular weight PEG. It has been shown that with qualitative analysis and model fitting of SANS data, the highest molecular weight tetra-functional PEG hydrogels have a remarkably homogeneous network structure with low junction functionality. However, there are still some small indications of inhomogeneity for the lowest molecular weight networks even in d-DMF, suggesting a higher level of defect formation during cross-linking for these systems. Despite the extensive research on structural study of materials using SANS and USANS, to date there has not been any reports of using a combination of SANS and USANS to study the effect of corrosion inhibitors' structure on their efficiency. Recently, using combined USANS/SANS tools, Taghavikish et al. [74] investigated the hierarchical gel network structure and their relation to the observed bulk properties for polymeric ionic liquid nanoparticle emulsion based corrosion inhibitor in anticorrosion coatings.

7. Conclusions and Future Outlook

The importance of physico-chemical properties of the emerging corrosion inhibitors based on IL and graphene for corrosion prevention of metallic substrates were outlined. Despite the very interesting properties of ILs and graphene for corrosion inhibition, there have been very few reports on their application for corrosion protection of metals. On the other hand, in spite of the considerable progress made in the chemistry of inhibitors, the inhibition study in a corrosion system presents the same challenges today as it did in the past. The transport of the corrosion inhibitor from bulk solution to the surface of the metal and the active inhibitor species that is available to interact with the metal are the fundamental factors governing corrosion inhibition. In the case of inhibitors that adsorb on the metal surface and inhibit the corrosion, there are two main challenges: (1) metal-inhibitor interactions and (2) transport of the corrosion inhibitor from bulk solution to the surface of the metal. These challenges could be overcome through immobilization of corrosion inhibitors into micro-/nanocontainers.

Although active corrosion inhibitors lead to sufficient temporary protection of the underlying metal, in the case of local damage of the protective coating, to reach an even more extended lifetime protection or to have repeatable healing at a specific location, an additional functionality of damage closure is required. Therefore, there is a significant need to develop the new micro-/nanocapsules family which are sensitive to pH, temperature, environmental conditions changes and dispersion of them to coatings should be considered.

Further research should also be focused on using more advanced characterization techniques and more fundamental studies in order to further clarify the mechanism of corrosion inhibitors and investigate the correlation between the structure and observed corrosion inhibition. This understanding will help us to tailor the inhibitor structure to obtain required corrosion inhibition properties. This would be more pronounced in the case of appropriate chemical modifications, especially by using eco-friendly inhibitors, which can render the coating with enhanced anticorrosive characteristics.

Acknowledgments: The authors gratefully acknowledge the financial support of the Australian Research Council's Linkage grant for carrying out this work and also the industry partner Wave rider Energy for financial support of this work.

Author Contributions: Namita Roy Choudhury, Mona Taghavikish and Naba Kumar Dutta designed the structure of the review. Mona Taghavikish wrote the review with Namita Roy Choudhury, Naba Kumar Dutta and the manuscript was finalized through contributions of all authors. All authors have given approval to the final version of the manuscript.

Conflicts of Interest: The authors declare no conflict of interest.

References

1. Wazarkar, K.; Patil, D.; Rane, A.; Balgude, D.; Kathalewar, M.; Sabnis, A. Microencapsulation: An emerging technique in the modern coating industry. *RSC Adv.* **2016**, *6*, 106964–106979. [CrossRef]
2. Dariva, C.G.; Galio, A.F. Corrosion inhibitors–principles, mechanisms and applications. In *Developments in Corrosion Protection*; Aliofkhazraei, M., Ed.; INTECH: Winchester, UK, 2014; pp. 365–379.
3. Benali, O.; Cherkaoui, O.; Lallam, A. Adsorption and corrosion inhibition of new synthesized pyridazinium-based ionic liquid on carbon steel in 0.5MH$_2$SO$_4$. *J. Mater. Environ. Sci.* **2015**, *6*, 598–606.
4. Saji, V.S. A review on recent patents in corrosion inhibitors. *Recent Pat. Corros. Sci.* **2010**, *2*, 6–12. [CrossRef]
5. Bardal, E. *Corrosion and Protection*; Springer Science & Business Media: London, UK, 2007.
6. Yaro, A.S.; Khadom, A.A.; Wael, R.K. Apricot juice as green corrosion inhibitor of mild steel in phosphoric acid. *Alex. Eng. J.* **2013**, *52*, 129–135. [CrossRef]
7. Sherif, E.-S.M. Effects of 2-amino-5-(ethylthio)-1,3,4-thiadiazole on copper corrosion as a corrosion inhibitor in 3% NaCl solutions. *Appl. Surf. Sci.* **2006**, *252*, 8615–8623. [CrossRef]
8. Revie, R.W. *Corrosion and Corrosion Control*, 4th ed.; John Wiley & Sons: Hoboken, NJ, USA, 2008.
9. Olivares, O.; Likhanova, N.; Gomez, B.; Navarrete, J.; Llanos-Serrano, M.; Arce, E.; Hallen, J. Electrochemical and XPS studies of decylamides of α-amino acids adsorption on carbon steel in acidic environment. *Appl. Surf. Sci.* **2006**, *252*, 2894–2909. [CrossRef]
10. Olivares-Xometl, O.; Likhanova, N.; Domínguez-Aguilar, M.; Arce, E.; Dorantes, H.; Arellanes-Lozada, P. Synthesis and corrosion inhibition of α-amino acids alkylamides for mild steel in acidic environment. *Mater. Chem. Phys.* **2008**, *110*, 344–351. [CrossRef]
11. El-Maksoud, S.A.; Fouda, A. Some pyridine derivatives as corrosion inhibitors for carbon steel in acidic medium. *Mater. Chem. Phys.* **2005**, *93*, 84–90. [CrossRef]
12. Noor, E.A. Evaluation of inhibitive action of some quaternary N-heterocyclic compounds on the corrosion of Al–Cu alloy in hydrochloric acid. *Mater. Chem. Phys.* **2009**, *114*, 533–541. [CrossRef]
13. Martínez-Palou, R.; Rivera, J.; Zepeda, L.; Rodríguez, A.; Hernández, M.; Marín-Cruz, J.; Estrada, A. Evaluation of corrosion inhibitors synthesized from fatty acids and fatty alcohols isolated from sugar cane wax. *Corrosion* **2004**, *60*, 465–470. [CrossRef]
14. Olivares-Xometl, O.; Likhanova, N.; Martínez-Palou, R.; Domínguez-Aguilar, M. Electrochemistry and XPS study of an imidazoline as corrosion inhibitor of mild steel in an acidic environment. *Mater. Corros.* **2009**, *60*, 14–21. [CrossRef]
15. Popova, A.; Christov, M.; Zwetanova, A. Effect of the molecular structure on the inhibitor properties of azoles on mild steel corrosion in 1 m hydrochloric acid. *Corros. Sci.* **2007**, *49*, 2131–2143. [CrossRef]
16. Antonijević, M.M.; Milić, S.M.; Petrović, M.B. Films formed on copper surface in chloride media in the presence of azoles. *Corros. Sci.* **2009**, *51*, 1228–1237. [CrossRef]
17. Tallman, D.E.; Spinks, G.; Dominis, A.; Wallace, G.G. Electroactive conducting polymers for corrosion control. *J. Solid State Electrochem.* **2002**, *6*, 73–84. [CrossRef]
18. Hu, J.; Ji, Y.; Shi, Y.; Hui, F.; Duan, H.; Lanza, M. A review on the use of graphene as a protective coating against corrosion. *Ann. Mater. Sci. Eng.* **2014**, *1*, 16.
19. Martínez-Palou, R.; Sánche, P.F. *Perspectives of Ionic Liquids Applications for Clean Oilfield Technologies*; INTECH Open Access Publisher: Rijeka, Croatia, 2011.
20. Zhang, Q.; Hua, Y. Corrosion inhibition of mild steel by alkylimidazolium ionic liquids in hydrochloric acid. *Electrochim. Acta* **2009**, *54*, 1881–1887. [CrossRef]
21. Yuan, J.; Antonietti, M. Poly(ionic liquid)s: Polymers expanding classical property profiles. *Polymer* **2011**, *52*, 1469–1482. [CrossRef]
22. Plechkova, N.V.; Seddon, K.R. Applications of ionic liquids in the chemical industry. *Chem. Soc. Rev.* **2008**, *37*, 123–150. [CrossRef] [PubMed]
23. Olivier-Bourbigou, H.; Magna, L.; Morvan, D. Ionic liquids and catalysis: Recent progress from knowledge to applications. *Appl. Catal. A General* **2010**, *373*, 1–56. [CrossRef]
24. Han, X.; Armstrong, D.W. Ionic liquids in separations. *Acc. Chem. Res.* **2007**, *40*, 1079–1086. [CrossRef] [PubMed]

25. Bara, J.E.; Carlisle, T.K.; Gabriel, C.J.; Camper, D.; Finotello, A.; Gin, D.L.; Noble, R.D. Guide to CO_2 separations in imidazolium-based room-temperature ionic liquids. *Ind. Eng. Chem. Res.* **2009**, *48*, 2739–2751. [CrossRef]
26. Anderson, J.L.; Armstrong, D.W.; Wei, G.-T. Ionic liquids in analytical chemistry. *Anal. Chem.* **2006**, *78*, 2892–2902. [CrossRef] [PubMed]
27. Wang, H.; Lu, Q.; Ye, C.; Liu, W.; Cui, Z. Friction and wear behaviors of ionic liquid of alkylimidazolium hexafluorophosphates as lubricants for steel/steel contact. *Wear* **2004**, *256*, 44–48. [CrossRef]
28. Kuang, D.; Wang, P.; Ito, S.; Zakeeruddin, S.M.; Grätzel, M. Stable mesoscopic dye-sensitized solar cells based on tetracyanoborate ionic liquid electrolyte. *J. Am. Chem. Soc.* **2006**, *128*, 7732–7733. [CrossRef] [PubMed]
29. Likhanova, N.V.; Domínguez-Aguilar, M.A.; Olivares-Xometl, O.; Nava-Entzana, N.; Arce, E.; Dorantes, H. The effect of ionic liquids with imidazolium and pyridinium cations on the corrosion inhibition of mild steel in acidic environment. *Corros. Sci.* **2010**, *52*, 2088–2097. [CrossRef]
30. Łuczak, J.; Hupka, J.; Thöming, J.; Jungnickel, C. Self-organization of imidazolium ionic liquids in aqueous solution. *Colloids Surf. A Physicochem. Eng. Asp.* **2008**, *329*, 125–133. [CrossRef]
31. Espinosa, T.; Sanes, J.; Jiménez, A.-E.; Bermúdez, M.-D. Surface interactions, corrosion processes and lubricating performance of protic and aprotic ionic liquids with ofhc copper. *Appl. Surf. Sci.* **2013**, *273*, 578–597. [CrossRef]
32. Elachouri, M.; Hajji, M.; Kertit, S.; Essassi, E.; Salem, M.; Coudert, R. Some surfactants in the series of 2-(alkyldimethylammonio) alkanol bromides as inhibitors of the corrosion of iron in acid chloride solution. *Corros. Sci.* **1995**, *37*, 381–389. [CrossRef]
33. Quraishi, M.A.; Rafiquee, M.Z.A.; Khan, S.; Saxena, N. Corrosion inhibition of aluminium in acid solutions by some imidazoline derivatives. *J. Appl. Electrochem.* **2007**, *37*, 1153–1162. [CrossRef]
34. Gasparac, R.; Martin, C.; Stupnisek-Lisac, E. In situ studies of imidazole and its derivatives as copper corrosion inhibitors. I. Activation energies and thermodynamics of adsorption. *J. Electrochem. Soc.* **2000**, *147*, 548–551. [CrossRef]
35. Zhang, D.-Q.; Gao, L.-X.; Zhou, G.-D. Inhibition of copper corrosion by bis-(1,1′-benzotriazoly)-α,ω-diamide compounds in aerated sulfuric acid solution. *Appl. Surf. Sci.* **2006**, *252*, 4975–4981. [CrossRef]
36. Zhang, Q.; Hua, Y. Corrosion inhibition of aluminum in hydrochloric acid solution by alkylimidazolium ionic liquids. *Mater. Chem. Phys.* **2010**, *119*, 57–64. [CrossRef]
37. Shuncun, S.; Pinggui, Y.; Chenzhong, C.; Xueye, W.; Jieshu, S.; Junxi, L. Synthesis of new ionic liquids and corrosion inhibition performance of its cationic imidazoline group. *J. Chem. Ind. Eng. China* **2005**, *56*, 1112–1119.
38. Ford, F.; Burstein, G.; Hoar, T. Bare surface reaction rates and their relation to environment controlled cracking of aluminum alloys I. Bare surface reaction rates on aluminum-7 weight percent magnesium in aqueous solutions. *J. Electrochem. Soc.* **1980**, *127*, 1325–1331. [CrossRef]
39. Uerdingen, M.; Treber, C.; Balser, M.; Schmitt, G.; Werner, C. Corrosion behaviour of ionic liquids. *Green Chem.* **2005**, *7*, 321–325. [CrossRef]
40. Ashassi-Sorkhabi, H.; Es'haghi, M. Corrosion inhibition of mild steel in acidic media by [BMIm]Br ionic liquid. *Mater. Chem. Phys.* **2009**, *114*, 267–271. [CrossRef]
41. Chong, A.L.; Mardel, J.I.; MacFarlane, D.R.; Forsyth, M.; Somers, A.E. Synergistic corrosion inhibition of mild steel in aqueous chloride solutions by an imidazolinium carboxylate salt. *ACS Sustain. Chem. Eng.* **2016**, *4*, 1746–1755. [CrossRef]
42. Zhang, H.; Hong, K.; Mays, J.W. Synthesis of block copolymers of styrene and methyl methacrylate by conventional free radical polymerization in room temperature ionic liquids. *Macromolecules* **2002**, *35*, 5738–5741. [CrossRef]
43. Woecht, I.; Schmidt-Naake, G.; Beuermann, S.; Buback, M.; García, N. Propagation kinetics of free-radical polymerizations in ionic liquids. *J. Polym. Sci. Part A Polym. Chem.* **2008**, *46*, 1460–1469. [CrossRef]
44. Ricks-Laskoski, H.L.; Snow, A.W. Synthesis and electric field actuation of an ionic liquid polymer. *J. Am. Chem. Soc.* **2006**, *128*, 12402–12403. [CrossRef] [PubMed]
45. Green, O.; Grubjesic, S.; Lee, S.; Firestone, M.A. The design of polymeric ionic liquids for the preparation of functional materials. *Polym. Rev.* **2009**, *49*, 339–360. [CrossRef]
46. Anderson, E.B.; Long, T.E. Imidazole- and imidazolium-containing polymers for biology and material science applications. *Polymer* **2010**, *51*, 2447–2454. [CrossRef]

47. Green, M.D.; Long, T.E. Designing imidazole-based ionic liquids and ionic liquid monomers for emerging technologies. *Polym. Rev.* **2009**, *49*, 291–314. [CrossRef]
48. Shaplov, A.S.; Vlasov, P.S.; Lozinskaya, E.I.; Shishkan, O.A.; Ponkratov, D.O.; Malyshkina, I.A.; Vidal, F.; Wandrey, C.; Godovikov, I.A.; Vygodskii, Y.S. Thiol-ene click chemistry as a tool for a novel family of polymeric ionic liquids. *Macromol. Chem. Phys.* **2012**, *213*, 1359–1369. [CrossRef]
49. Shaplov, A.S.; Goujon, L.; Vidal, F.; Lozinskaya, E.I.; Meyer, F.; Malyshkina, I.A.; Chevrot, C.; Teyssie, D.; Odinets, I.L.; Vygodskii, Y.S. Ionic IPNs as novel candidates for highly conductive solid polymer electrolytes. *J. Polym. Sci. Part A Polym. Chem.* **2009**, *47*, 4245–4266. [CrossRef]
50. Allen, M.H.; Green, M.D.; Getaneh, H.K.; Miller, K.M.; Long, T.E. Tailoring charge density and hydrogen bonding of imidazolium copolymers for efficient gene delivery. *Biomacromolecules* **2011**, *12*, 2243–2250. [CrossRef] [PubMed]
51. Qiu, B.; Lin, B.; Si, Z.; Qiu, L.; Chu, F.; Zhao, J.; Yan, F. Bis-imidazolium-based anion-exchange membranes for alkaline fuel cells. *J. Power Sources* **2012**, *217*, 329–335. [CrossRef]
52. Mecerreyes, D. Polymeric ionic liquids: Broadening the properties and applications of polyelectrolytes. *Prog. Polym. Sci.* **2011**, *36*, 1629–1648. [CrossRef]
53. Arellanes-Lozada, P.; Olivares-Xometl, O.; Guzmán-Lucero, D.; Likhanova, N.V.; Domínguez-Aguilar, M.A.; Lijanova, I.V.; Arce-Estrada, E. The inhibition of aluminum corrosion in sulfuric acid by poly (1-vinyl-3-alkyl-imidazolium hexafluorophosphate). *Materials* **2014**, *7*, 5711–5734. [CrossRef] [PubMed]
54. Atta, A.M.; El-Mahdy, G.A.; Allohedan, H.A.; Abdullah, M.M. Poly (ionic liquid) based on modified ionic polyacrylamide for inhibition steel corrosion in acid solution. *Int. J. Electrochem. Sci.* **2015**, *10*, 10389–10401.
55. Borisova, D.; Akçakayıran, D.; Schenderlein, M.; Möhwald, H.; Shchukin, D.G. Nanocontainer-based anticorrosive coatings: Effect of the container size on the self-healing performance. *Adv. Funct. Mater.* **2013**, *23*, 3799–3812. [CrossRef]
56. Abdelhedi-Miladi, I.; Obadia, M.M.; Allaoua, I.; Serghei, A.; Romdhane, H.B.; Drockenmuller, E. 1,2,3-triazolium-based poly(ionic liquid)s obtained through click chemistry polyaddition. *Macromol. Chem. Phys.* **2014**, *215*, 2229–2236. [CrossRef]
57. Stepto Robert, F.T. Dispersity in polymer science (IUPAC recommendations 2009). *Pure Appl. Chem.* **2009**, *81*, 351–353. [CrossRef]
58. Muldoon, M.J.; Gordon, C.M. Synthesis of gel-type polymer beads from ionic liquid monomers. *J. Polym. Sci. Part A Polym. Chem.* **2004**, *42*, 3865–3869. [CrossRef]
59. Marcilla, R.; Sanchez-Paniagua, M.; Lopez-Ruiz, B.; Lopez-Cabarcos, E.; Ochoteco, E.; Grande, H.; Mecerreyes, D. Synthesis and characterization of new polymeric ionic liquid microgels. *J. Polym. Sci. Part A Polym. Chem.* **2006**, *44*, 3958–3965. [CrossRef]
60. Yuan, J.; Antonietti, M. Poly(ionic liquid) latexes prepared by dispersion polymerization of ionic liquid monomers. *Macromolecules* **2011**, *44*, 744–750. [CrossRef]
61. Yuan, J.; Soll, S.; Drechsler, M.; Müller, A.H.E.; Antonietti, M. Self-assembly of poly(ionic liquid)s: Polymerization, mesostructure formation, and directional alignment in one step. *J. Am. Chem. Soc.* **2011**, *133*, 17556–17559. [CrossRef] [PubMed]
62. Weber, R.L.; Ye, Y.; Schmitt, A.L.; Banik, S.M.; Elabd, Y.A.; Mahanthappa, M.K. Effect of nanoscale morphology on the conductivity of polymerized ionic liquid block copolymers. *Macromolecules* **2011**, *44*, 5727–5735. [CrossRef]
63. Yang, J.; Qiu, L.; Liu, B.; Peng, Y.; Yan, F.; Shang, S. Synthesis of polymeric ionic liquid microsphere/Pt nanoparticle hybrids for electrocatalytic oxidation of methanol and catalytic oxidation of benzyl alcohol. *J. Polym. Sci. Part A Polym. Chem.* **2011**, *49*, 4531–4538. [CrossRef]
64. Zhou, X.; Zhou, Y.; Nie, J.; Ji, Z.; Xu, J.; Zhang, X.; Du, B. Thermosensitive ionic microgels via surfactant-free emulsion copolymerization and in situ quaternization cross-linking. *ACS Appl. Mater. Interfaces* **2014**, *6*, 4498–4513. [CrossRef] [PubMed]
65. Amato, D.V.; Amato, D.N.; Flynt, A.S.; Patton, D.L. Functional, sub-100 nm polymer nanoparticles via thiol-ene miniemulsion photopolymerization. *Polym. Chem.* **2015**, *6*, 5625–5632. [CrossRef]
66. Hoyle, C.E.; Lee, T.Y.; Roper, T. Thiol–enes: Chemistry of the past with promise for the future. *J. Polym. Sci. Part A Polym. Chem.* **2004**, *42*, 5301–5338. [CrossRef]
67. Hoyle, C.E.; Lowe, A.B.; Bowman, C.N. Thiol-click chemistry: A multifaceted toolbox for small molecule and polymer synthesis. *Chem. Soc. Rev.* **2010**, *39*, 1355–1387. [CrossRef] [PubMed]

68. Jasinski, F.; Lobry, E.; Tarablsi, B.; Chemtob, A.; Croutxé-Barghorn, C.; Le Nouen, D.; Criqui, A. Light-mediated thiol–ene polymerization in miniemulsion: A fast route to semicrystalline polysulfide nanoparticles. *ACS Macro Lett.* **2014**, *3*, 958–962. [CrossRef]
69. Taghavikish, M.; Subianto, S.; Dutta, N.K.; de Campo, L.; Mata, J.P.; Rehm, C.; Choudhury, N.R. Polymeric ionic liquid nanoparticle emulsions as a corrosion inhibitor in anticorrosion coatings. *ACS Omega* **2016**, *1*, 29–40. [CrossRef]
70. Raghavan, P.; Manuel, J.; Zhao, X.; Kim, D.-S.; Ahn, J.-H.; Nah, C. Preparation and electrochemical characterization of gel polymer electrolyte based on electrospun polyacrylonitrile nonwoven membranes for lithium batteries. *J. Power Sources* **2011**, *196*, 6742–6749. [CrossRef]
71. Murata, Y.; Sasaki, N.; Miyamoto, E.; Kawashima, S. Use of floating alginate gel beads for stomach-specific drug delivery. *Eur. J. Pharm. Biopharm.* **2000**, *50*, 221–226. [CrossRef]
72. Kashyap, N.; Kumar, N.; Kumar, M.N.V.R. Hydrogels for pharmaceutical and biomedical applications. *Crit. Rev. Ther. Drug Carr. Syst.* **2005**, *22*, 107–149. [CrossRef]
73. Xiong, Y.; Wang, H.; Wang, R.; Yan, Y.; Zheng, B.; Wang, Y. A facile one-step synthesis to cross-linked polymeric nanoparticles as highly active and selective catalysts for cycloaddition of CO_2 to epoxides. *Chem. Commun.* **2010**, *46*, 3399–3401. [CrossRef] [PubMed]
74. Rahman, M.T.; Barikbin, Z.; Badruddoza, A.Z.M.; Doyle, P.S.; Khan, S.A. Monodisperse polymeric ionic liquid microgel beads with multiple chemically switchable functionalities. *Langmuir* **2013**, *29*, 9535–9543. [CrossRef] [PubMed]
75. Taghavikish, M.; Subianto, S.; Dutta, N.K.; Choudhury, N.R. Facile fabrication of polymerizable ionic liquid based-gel beads via thiol–ene chemistry. *ACS Appl. Mater. Interfaces* **2015**, *7*, 17298–17306. [CrossRef] [PubMed]
76. Graphene. Available online: https://www.graphene-info.com/introduction (accessed on 24 October 2017).
77. Prasai, D.; Tuberquia, J.C.; Harl, R.R.; Jennings, G.K.; Bolotin, K.I. Graphene: Corrosion-inhibiting coating. *ACS Nano* **2012**, *6*, 1102–1108. [CrossRef] [PubMed]
78. Zhu, Y.; Murali, S.; Cai, W.; Li, X.; Suk, J.W.; Potts, J.R.; Ruoff, R.S. Graphene and graphene oxide: Synthesis, properties, and applications. *Adv. Mater.* **2010**, *22*, 3906–3924. [CrossRef] [PubMed]
79. Qiu, J.; Wang, S. Enhancing polymer performance through graphene sheets. *J. Appl. Polym. Sci.* **2011**, *119*, 3670–3674. [CrossRef]
80. Lee, C.; Wei, X.; Kysar, J.W.; Hone, J. Measurement of the elastic properties and intrinsic strength of monolayer graphene. *Science* **2008**, *321*, 385–388. [CrossRef] [PubMed]
81. Balandin, A.A.; Ghosh, S.; Bao, W.; Calizo, I.; Teweldebrhan, D.; Miao, F.; Lau, C.N. Superior thermal conductivity of single-layer graphene. *Nano Lett.* **2008**, *8*, 902–907. [CrossRef] [PubMed]
82. Cheng, J.; Zhang, G.; Du, J.; Tang, L.; Xu, J.; Li, J. New role of graphene oxide as active hydrogen donor in the recyclable palladium nanoparticles catalyzed ullmann reaction in environmental friendly ionic liquid/supercritical carbon dioxide system. *J. Mater. Chem.* **2011**, *21*, 3485–3494. [CrossRef]
83. Song, H.; Zhang, L.; He, C.; Qu, Y.; Tian, Y.; Lv, Y. Graphene sheets decorated with SnO_2 nanoparticles: In situ synthesis and highly efficient materials for cataluminescence gas sensors. *J. Mater. Chem.* **2011**, *21*, 5972–5977. [CrossRef]
84. Zhao, J.; Pei, S.; Ren, W.; Gao, L.; Cheng, H.-M. Efficient preparation of large-area graphene oxide sheets for transparent conductive films. *ACS Nano* **2010**, *4*, 5245–5252. [CrossRef] [PubMed]
85. Liang, M.; Zhi, L. Graphene-based electrode materials for rechargeable lithium batteries. *J. Mater. Chem.* **2009**, *19*, 5871–5878. [CrossRef]
86. Reddy, A.L.M.; Srivastava, A.; Gowda, S.R.; Gullapalli, H.; Dubey, M.; Ajayan, P.M. Synthesis of nitrogen-doped graphene films for lithium battery application. *ACS Nano* **2010**, *4*, 6337–6342. [CrossRef] [PubMed]
87. Zhang, K.; Zhang, L.L.; Zhao, X.S.; Wu, J. Graphene/polyaniline nanofiber composites as supercapacitor electrodes. *Chem. Mater.* **2010**, *22*, 1392–1401. [CrossRef]
88. Kalita, G.; Matsushima, M.; Uchida, H.; Wakita, K.; Umeno, M. Graphene constructed carbon thin films as transparent electrodes for solar cell applications. *J. Mater. Chem.* **2010**, *20*, 9713–9717. [CrossRef]
89. Ramanathan, T.; Abdala, A.A.; Stankovich, S.; Dikin, D.A.; Herrera Alonso, M.; Piner, R.D.; Adamson, D.H.; Schniepp, H.C.; Chen, X.; Ruoff, R.S.; et al. Functionalized graphene sheets for polymer nanocomposites. *Nat. Nanotechnol.* **2008**, *3*, 327–331. [CrossRef] [PubMed]

90. Salavagione, H.J.; Martinez, G.; Gomez, M.A. Synthesis of poly(vinyl alcohol)/reduced graphite oxide nanocomposites with improved thermal and electrical properties. *J. Mater. Chem.* **2009**, *19*, 5027–5032. [CrossRef]
91. Cai, D.; Song, M. Recent advance in functionalized graphene/polymer nanocomposites. *J. Mater. Chem.* **2010**, *20*, 7906–7915. [CrossRef]
92. Zhu, J. Graphene production: New solutions to a new problem. *Nat Nano* **2008**, *3*, 528–529. [CrossRef] [PubMed]
93. Chang, C.-H.; Huang, T.-C.; Peng, C.-W.; Yeh, T.-C.; Lu, H.-I.; Hung, W.-I.; Weng, C.-J.; Yang, T.-I.; Yeh, J.-M. Novel anticorrosion coatings prepared from polyaniline/graphene composites. *Carbon* **2012**, *50*, 5044–5051. [CrossRef]
94. Yasmin, A.; Daniel, I.M. Mechanical and thermal properties of graphite platelet/epoxy composites. *Polymer* **2004**, *45*, 8211–8219. [CrossRef]
95. Rafiee, M.A.; Rafiee, J.; Wang, Z.; Song, H.; Yu, Z.Z.; Koratkar, N. Enhanced mechanical properties of nanocomposites at low graphene content. *ACS Nano* **2009**, *3*, 3884–3890. [CrossRef] [PubMed]
96. Miller, S.G.; Bauer, J.L.; Maryanski, M.J.; Heimann, P.J.; Barlow, J.P.; Gosau, J.M.; Allred, R.E. Characterization of epoxy functionalized graphite nanoparticles and the physical properties of epoxy matrix nanocomposites. *Compos. Sci. Technol.* **2010**, *70*, 1120–1125. [CrossRef]
97. Chiang, C.L.; Hsu, S.W. Synthesis, characterization and thermal properties of novel epoxy/expandable graphite composites. *Polym. Int.* **2010**, *59*, 119–126. [CrossRef]
98. Martin-Gallego, M.; Verdejo, R.; Lopez-Manchado, M.A.; Sangermano, M. Epoxy-graphene UV-cured nanocomposites. *Polymer* **2011**, *52*, 4664–4669. [CrossRef]
99. Sørensen, P.A.; Kiil, S.; Dam-Johansen, K.; Weinell, C.E. Anticorrosive coatings: A review. *J. Coat. Technol. Res.* **2009**, *6*, 135–176. [CrossRef]
100. Bao, C.; Guo, Y.; Song, L.; Kan, Y.; Qian, X.; Hu, Y. In situ preparation of functionalized graphene oxide/epoxy nanocomposites with effective reinforcements. *J. Mater. Chem.* **2011**, *21*, 13290–13298. [CrossRef]
101. Borisova, D.; Möhwald, H.; Shchukin, D.G. Mesoporous silica nanoparticles for active corrosion protection. *ACS Nano* **2011**, *5*, 1939–1946. [CrossRef] [PubMed]
102. Shchukin, D.G.; Grigoriev, D.O.; Mohwald, H. Application of smart organic nanocontainers in feedback active coatings. *Soft Matter* **2010**, *6*, 720–725. [CrossRef]
103. Dry, C. Procedures developed for self-repair of polymer matrix composite materials. *Compos. Struct.* **1996**, *35*, 263–269. [CrossRef]
104. Dry, C.M.; Sottos, N.R. Passive smart self-repair in polymer matrix composite materials. *Proc. SPIE Int. Soc. Opt. Eng.* **1993**, *1916*, 438–444.
105. Zhang, M.Q.; Rong, M.Z. Intrinsic self-healing of covalent polymers through bond reconnection towards strength restoration. *Polym. Chem.* **2013**, *4*, 4878–4884. [CrossRef]
106. Chen, Y.; Kushner, A.M.; Williams, G.A.; Guan, Z. Multiphase design of autonomic self-healing thermoplastic elastomers. *Nat. Chem.* **2012**, *4*, 467–472. [CrossRef] [PubMed]
107. Kessler, M.R. Self-healing: A new paradigm in materials design. *Proc. Inst. Mech. Eng. Part G J. Aerosp. Eng.* **2007**, *221*, 479–495. [CrossRef]
108. Lvov, Y.M.; Shchukin, D.G.; Möhwald, H.; Price, R.R. Halloysite clay nanotubes for controlled release of protective agents. *ACS Nano* **2008**, *2*, 814–820. [CrossRef] [PubMed]
109. Agnihotri, N.; Mishra, R.; Goda, C.; Arora, M. Microencapsulation—A novel approach in drug delivery: A review. *Indo Glob. J. Pharm. Sci.* **2012**, *2*, 1–20.
110. Liu, X.; Zhang, H.; Wang, J.; Wang, Z.; Wang, S. Preparation of epoxy microcapsule based self-healing coatings and their behavior. *Surf. Coat. Technol.* **2012**, *206*, 4976–4980. [CrossRef]
111. Huang, M.; Zhang, H.; Yang, J. Synthesis of organic silane microcapsules for self-healing corrosion resistant polymer coatings. *Corros. Sci.* **2012**, *65*, 561–566. [CrossRef]
112. Brown, E.N.; Kessler, M.R.; Sottos, N.R.; White, S.R. In situ poly(urea-formaldehyde) microencapsulation of dicyclopentadiene. *J. Microencapsul.* **2003**, *20*, 719–730. [CrossRef] [PubMed]
113. Cho, S.H.; White, S.R.; Braun, P.V. Self-healing polymer coatings. *Adv. Mater.* **2009**, *21*, 645–649. [CrossRef]
114. Samadzadeh, M.; Boura, S.H.; Peikari, M.; Kasiriha, S.M.; Ashrafi, A. A review on self-healing coatings based on micro/nanocapsules. *Prog. Org. Coat.* **2010**, *68*, 159–164. [CrossRef]

115. Wilson, G.O.; Andersson, H.M.; White, S.R.; Sottos, N.R.; Moore, J.S.; Braun, P.V. Self-healing polymers. In *Encyclopedia of Polymer Science and Technology*; John Wiley & Sons, Inc.: New York, NY, USA, 2002.
116. Bandeira, P.; Monteiro, J.; Baptista, A.M.; Magalhães, F.D. Tribological performance of PTFE-based coating modified with microencapsulated [HMIM][NTf2] ionic liquid. *Tribol. Lett.* **2015**, *59*, 1–15. [CrossRef]
117. García, S.J.; Fischer, H.R.; White, P.A.; Mardel, J.; González-García, Y.; Mol, J.M.C.; Hughes, A.E. Self-healing anticorrosive organic coating based on an encapsulated water reactive silyl ester: Synthesis and proof of concept. *Prog. Org. Coat.* **2011**, *70*, 142–149. [CrossRef]
118. Minami, H.; Fukaumi, H.; Okubo, M.; Suzuki, T. Preparation of ionic liquid-encapsulated polymer particles. *Colloid Polym. Sci.* **2013**, *291*, 45–51. [CrossRef]
119. Minami, H.; Okubo, M.; Oshima, Y. Preparation of cured epoxy resin particles having one hollow by polyaddition reaction. *Polymer* **2005**, *46*, 1051–1056. [CrossRef]
120. Nesterova, T.; Dam-Johansen, K.; Pedersen, L.T.; Kiil, S. Microcapsule-based self-healing anticorrosive coatings: Capsule size, coating formulation, and exposure testing. *Prog. Org. Coat.* **2012**, *75*, 309–318. [CrossRef]
121. Mehta, N.K.; Bogere, M.N. Environmental studies of smart/self-healing coating system for steel. *Prog. Org. Coat.* **2009**, *64*, 419–428. [CrossRef]
122. Kumar, A.; Mathur, N. Photocatalytic degradation of aniline at the interface of TiO_2 suspensions containing carbonate ions. *J. Colloid Interface Sci.* **2006**, *300*, 244–252. [CrossRef] [PubMed]
123. Otmacic Curkovic, H.; Stupnisek-Lisac, E.; Takenouti, H. Electrochemical quartz crystal microbalance and electrochemical impedance spectroscopy study of copper corrosion inhibition by imidazoles. *Corros. Sci.* **2009**, *51*, 2342–2348. [CrossRef]
124. Dornbusch, M. The use of modern electrochemical methods in the development of corrosion protective coatings. *Prog. Org. Coat.* **2008**, *61*, 240–244. [CrossRef]
125. Nazarov, A.; Prosek, T.; Thierry, D. Application of EIS and SKP methods for the study of the zinc/polymer interface. *Electrochim. Acta* **2008**, *53*, 7531–7538. [CrossRef]
126. Wapner, K.; Stratmann, M.; Grundmeier, G. Application of the scanning kelvin probe for the study of the corrosion resistance of interfacial thin organosilane films at adhesive/metal interfaces. *Silicon Chem.* **2005**, *2*, 235–245. [CrossRef]
127. Kannan, A.G.; Choudhury, N.R.; Dutta, N.K. Electrochemical performance of sol-gel derived phospho-silicate-methacrylate hybrid coatings. *J. Electroanal. Chem.* **2010**, *641*, 28–34. [CrossRef]
128. Kannan, A.G.; Choudhury, N.R.; Dutta, N.K. Synthesis and characterization of methacrylate phospho-silicate hybrid for thin film applications. *Polymer* **2007**, *48*, 7078–7086. [CrossRef]
129. Stratmann, M. The investigation of the corrosion properties of metals, covered with adsorbed electrolyte layers—A new experimental technique. *Corros. Sci.* **1987**, *27*, 869–872. [CrossRef]
130. Pepe, A.; Galliano, P.; Aparicio, M.; Durán, A.; Ceré, S. Sol-gel coatings on carbon steel: Electrochemical evaluation. *Surf. Coat. Technol.* **2006**, *200*, 3486–3491. [CrossRef]
131. Sherif, E.-S.M.; Erasmus, R.M.; Comins, J.D. In situ raman spectroscopy and electrochemical techniques for studying corrosion and corrosion inhibition of iron in sodium chloride solutions. *Electrochim. Acta* **2010**, *55*, 3657–3663. [CrossRef]
132. Shibayama, M. Small-angle neutron scattering on polymer gels: Phase behavior, inhomogeneities and deformation mechanisms. *Polym. J.* **2011**, *43*, 18–34. [CrossRef]
133. Matsunaga, T.; Sakai, T.; Akagi, Y.; Chung, U.-I.; Shibayama, M. SANS and SLS studies on Tetra-Arm PEG gels in as-prepared and swollen states. *Macromolecules* **2009**, *42*, 6245–6252. [CrossRef]
134. Saffer, E.M.; Lackey, M.A.; Griffin, D.M.; Kishore, S.; Tew, G.N.; Bhatia, S.R. Sans study of highly resilient poly (ethylene glycol) hydrogels. *Soft Matter* **2014**, *10*, 1905–1916. [CrossRef] [PubMed]

© 2017 by the authors. Licensee MDPI, Basel, Switzerland. This article is an open access article distributed under the terms and conditions of the Creative Commons Attribution (CC BY) license (http://creativecommons.org/licenses/by/4.0/).

Article

Evolution of the Three-Dimensional Structure and Growth Model of Plasma Electrolytic Oxidation Coatings on 1060 Aluminum Alloy

Xiaohui Liu [1], Shuaixing Wang [1,*], Nan Du [1,*], Xinyi Li [2] and Qing Zhao [1]

1. National Defense Key Discipline Laboratory of Light Alloy Processing Science and Technology, Nanchang Hangkong University, Nanchang 330063, China; xiaohuiliu612@163.com (X.L.); z_haoqing@sina.com (Q.Z.)
2. Corrosion and Protection Center, University of Science and Technology Beijing, Beijing 100083, China; xylnchu@163.com
* Correspondence: wsxxpg@126.com (S.W.); d_unan@sina.com (N.D.); Tel.: +86-791-8386-3187 (S.W. & N.D.)

Received: 28 January 2018; Accepted: 13 March 2018; Published: 15 March 2018

Abstract: A deeper understanding of plasma electrolytic oxidation (PEO) can in turn shed light on the evolution of coating structures during such oxidation processes. Here, a three-dimensional (3D) structure of PEO coating was investigated based on the morphologies at different locations in a PEO coating and on the elemental distribution along certain sections. The coating surface was dominated by a crater- or pancake-like structure of alumina surrounded by Si-rich nodules. A barrier layer with a thickness of ~1 μm consisting of clustered cells was present at the aluminum/coating interface. As the coating thickened, the PEO coating gradually evolved into a distinct three-layer structure, which included a barrier layer, an internal structure with numerous closed holes, and an outer layer with a rough surface. During the PEO process, molten zones formed along with the plasma discharges. The volume and lifetime of the molten zones changed with oxidation time. The diversities of cooling rates around the molten zones resulted in structural differences along a certain section of the coating. A growth and discharge model of PEO coatings was established based on the 3D structure of the particular coating studied herein.

Keywords: plasma electrolytic oxidation (PEO); aluminum; three-dimensional structure; aluminum/coating interface; growth model

1. Introduction

Plasma electrolytic oxidation (PEO) is a surface-modification technique for producing ceramic coatings on light metals and their alloys (such as aluminum, magnesium, and titanium) [1,2]. PEO coating is considered to be amongst the most promising protective coatings for application in a wide range of industry sectors because of its high microhardness and its good wear and corrosion resistance [3,4]. However, the long-term protection performance of a PEO coating is limited by its high porosity. Researchers have different opinions on whether the pores in the coating extend to the substrate [5,6]. Hence, the three-dimensional (3D) structure and growth mechanism of PEO coatings need to be studied.

In general, most information about the structure of PEO coatings is obtained from the conventional surface and polished cross section. The coating surface is porous and coarse, consisting of pancake-like structures with a central hole [7]. PEO coatings are divided into three layers, i.e., an outer loose layer, an inner dense layer, and the barrier layer near the substrate [8,9]. A free-standing coating detached from the substrate can be used to obtain more information about the PEO coating structure, e.g., the structure of the coating/substrate interface. Some researchers have tried to use chemical solutions to detach the coating from the aluminum substrate. However, chemical dissolution in NaOH may dissolve alumina coating [10,11], and chemical dissolution in $CuCl_2$ may lead to a copper cover at the coating surface [12]. Recently, free-standing coatings have been obtained via dissolving the coated aluminum with an electrochemical method [13,14]. Moreover, 3D information about the porosity of PEO coating structures has been obtained by X-ray computed tomography [6,15] and the resin replica method [16]. Additional information on the 3D structure of PEO coatings, especially the evolution of the 3D structure during the PEO process, is needed for a deeper understanding of the PEO mechanism.

For the growth of PEO coatings, the most commonly accepted mechanism is attributed to an outward–inward growth mechanism [17–21]. The presence of inward growth has been confirmed by ^{18}O element labeling [22], which is regarded as a process of repetitive breakdown and passivation of the barrier layer at the coating/substrate interface [13]. Additionally, the outward growth of PEO coatings has been shown by analyzing the elemental distribution in PEO coatings prepared on a substrate of Ti covering Al [19]. The outward growth of coatings is usually considered a process of ejecting molten oxide [18,23,24]. Another theory is that the outward growth of PEO coatings occurs because the outer layer expands outwards under a squeezing effect owing to a thickening barrier layer [25]. Further studies of the PEO mechanism are limited by a lack of understanding of the coating structure.

In this work, PEO coatings were prepared on 1060 aluminum alloy in the silicate-phosphate electrolyte. The free-standing coating was obtained by the dissolution of substrate using an electrochemical method. The 3D structure of the PEO coating was analyzed using a field emission gun SEM (FE-SEM) and energy dispersive spectroscopy (EDS) by layer-by-layer thinning. The 3D structure of the coating, including the surface, the internal structure, the aluminum/coating interface, and the fracture cross-section structure, was studied in detail. Based on the above results, a growth model of PEO coatings is proposed.

2. Materials and Methods

2.1. Materials and PEO Treatment

The 1060 aluminum alloy (99.6% purity) was used as substrate in this experiment. Specimens with the dimension of 20 mm × 20 mm × 2 mm were polished using silicon-carbide abrasive paper up to 1000 grit and cleaned with acetone for 5 min.

The PEO process was performed using a direct-current pulsed power supply (WHD-20, Harbin Institute of Technology, Harbin, China) at a constant current density of 15 A/dm^2, a frequency of 500 Hz, and a duty ratio of 60%. Electrolyte, mainly comprised of 8 g/L $Na_2SiO_3 \cdot 9H_2O$ and 3 g/L $Na_5P_3O_{10}$, was prepared by deionized water and high-purity chemicals. A water-cooling system was used to maintain an electrolyte temperature below 300 K. The specimens were subjected to PEO treatment for 5, 15, 45, and 60 min, cleaned ultrasonically in deionized water and dried in warm air.

2.2. Specimen Characterization

The 3D structure of PEO coating, including a surface, an internal structure, an aluminum/coating interface, and a fracture cross section, was obtained by an electrochemical dissolution method, a layer-by-layer thinning method, and a cross-cutting method. Among them, the internal structure was obtained by exfoliating the outer layer using a single scratch tester (WS-2005, Lanzhou Institute of Chemical Physics, CAS, Lanzhou, China). The scratch tests were made with a diamond indenter (120° cone with a tip with a radius of 200 μm) under a loading rate of 50 N/min, a loading range of 0–100 N, and a scratch length of 10 mm. The free-standing coating was obtained by removing the substrate using an electrochemical method [13,14]. The specific operation process is as follows: specimens with PEO coatings were firstly polished from one side to a ~0.5 mm thickness. The specimen was immersed partly into a 3.5 wt % NaCl solution as the anode and a stainless-steel cell as the cathode, which was connected with a direct-current power supply with an operating voltage of 15 V. The aluminum had been dissolved completely when the current decreased to zero. A free-standing coating was obtained, cleaned in deionized water, and left to dry naturally.

The surface, internal structure, coating/substrate interface, and fractured cross-sectional morphologies of the PEO coatings were observed by a field emission gun SEM (FE-SEM, Nova Nano SEM 450, Thermo Fisher Scientific, Hillsboro, OR, USA). Chemical compositions of the PEO coatings were analyzed by energy dispersive spectroscopy (EDS, INCA 250, Oxford Instruments, Oxford, UK) equipped on the FE-SEM system. The high voltage and spot size were set to 15 kV and 3.0, respectively. During EDS semi-quantitative analysis, five measurements were taken randomly, and the average value was taken. The standardless quantitative analysis and the XPP correction method were used for the atomic percent calculation of various elements during EDS analysis. For the elements investigated in this study (O, Al, Si, and P), the total relative uncertainty was estimated to be ±5%. The phase composition of the coatings was analyzed by X-ray diffraction (XRD, D8-Advance, Bruker, Karlsruhe, Germany) using Cu Kα radiation operated at 40 kV and 40 mA in a range from 10° to 80° with a step size of 0.02°. Before XRD analysis, free-standing coatings were obtained using the electrochemical method and then ground into powder samples using the mortar. The maximum height difference between the peaks and valleys of the coating was characterized by a 3D digital video microscope (KH-7700, Hirox Co., Ltd., Tokyo, Japan). The size of the 3D scanning region in this work was 295 μm × 220 μm. The 3D scanning resolution was 1 μm. The resolution of Z-axis step motor was 0.05 μm/pulse.

3. Results

3.1. Fracture Cross-Sectional Morphology and Chemical Composition of the PEO Coating

Free-standing coatings were obtained by electrochemically dissolving the aluminum substrate in the NaCl solution. Figure 1 shows the SEM images of the fracture cross-section of the free-standing PEO coatings and the corresponding EDS results at different treatment times. Figure 1a–d shows that all aluminum/coating interfaces had a wavy-jagged appearance, which may be a result of discontinuous oxidation of the aluminum substrate. A dense barrier layer with a relatively constant thickness of ~1 μm existed near the aluminum/coating interface.

In the initial stage (Figure 1a), the coating was compact with a thickness of ~1.2 μm. At 15 min (Figure 1b), the fracture cross-section displayed a clear longitudinal profile of the strip pores, which was thought to be the residue of the discharge channels in the PEO process. As shown in Figure 1c,d, the PEO coating with a three-layer structure was clearly revealed at 45–60 min, and large cavities were present in the coating. Discontinuous nodules were distributed over the outer surface. The internal structure was filled with a large number of micropores, and some cracks appeared to traverse the entire outer-layer thickness.

EDS analysis was conducted to achieve a better understanding of the cross-sectional structure. As shown in Figure 1e,f, the EDS point analysis (Point 4) revealed that the nodules were rich in Si. A small amount of Si was also detected at the edge of the cavity (Point 2). Very little P was detected at Point 2 (Figure 1d), since the oxides containing P were hard to deposit in the coatings [26]. It is noted that the PEO coatings contain a certain amount of Si and P in addition to aluminum, and these elements may also combine with oxygen, causing a higher O/Al ratio than expected. Electrolyte evaporation, condensation, decomposition, and deposition were caused by the heat of the plasma discharges, which resulted in the incorporation of the electrolyte composition into the PEO coating.

Figure 1g gives the typical EDS mappings corresponding to the cross section of the PEO coating (Figure 1d). It can be seen that Si was higher in nodules at the surface, suggesting that the silicate in the electrolyte was prone to be deposited to form Si-rich nodules at the surface of PEO coating. However, the P mostly distributed around the cavity, showing that the electrolyte had been penetrated into the cavity during the PEO process. The higher level of P in the inner coating might be related to the short-circuit transport of electrolyte components through the outer coating, which would be left inside the coating [27].

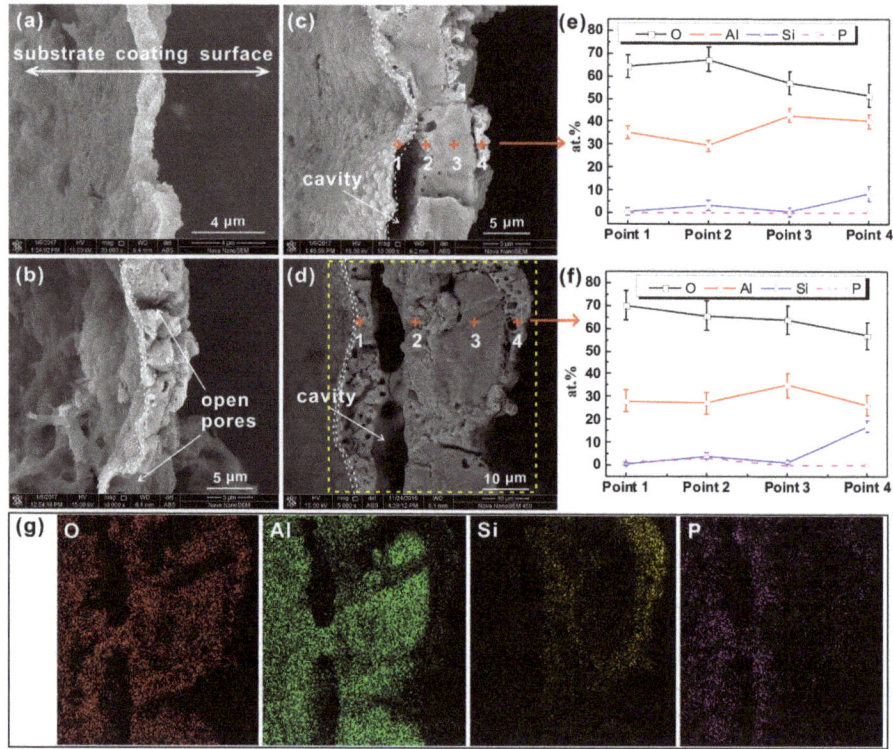

Figure 1. (**a–d**) SEM images of fracture cross section of PEO coatings formed at (**a**) 5 min, (**b**) 15 min, (**c**) 45 min, and (**d**) 60 min. (**e,f**) The corresponding EDS results to the coating of (**c**) and (**d**). (**g**) Typical EDS mappings of PEO coatings marked in Figure 1d.

3.2. Surface Morphology and Chemical Composition of PEO Coating

Figure 2 shows the SEM and 3D images at the surface of PEO coatings prepared at various oxidation times. As shown in Figure 2a–d, a large number of micropores and cracks were present at the surface of PEO coatings. Moreover, there were two categories for all coating surfaces. One is a loose nodule, and the other is a molten-shaped product with open or sealed micropores in the center. The morphology and size of the molten-shaped products changed with the increase in oxidation time. At 15 min, the molten-shaped products presented as craters with an open pore in the center. After 45 min, the molten-shaped products looked like pancakes with a sealed pore in the center. Additionally, 3D images displayed the evolution of the height difference at the coating surface (Figure 2e–h), indicating that nodules also enlarged during the PEO process.

In the initial stage (5 min), the PEO coating was compact and flat. As shown in Figure 2a,e, many fine nodules were present at the coating surface, but the molten-shaped product was not obvious. At 15 min, a large number of striped and sub-circular pores appeared at the PEO coating surface, and a group of nodules were present around these pores, as shown in Figure 2b. The 3D image (Figure 2f) also shows many obvious peaks and valleys. However, the surface morphology of the PEO coatings that formed at 45–60 min was obviously different from that of the thinner coatings. As shown in Figure 2c,d, pancake-like structures with a central pore were the main feature of these coatings, while larger nodules were also present. In addition, the height difference at the coating surface increased slightly as the oxidation increased from 45 to 60 min, which mainly resulted from the changes in nodules.

The elemental content of the PEO coatings were investigated by EDS analysis marked in Figure 2, as presented in Table 1. All coatings were composed of O, Al, and Si. Al in the coatings originated from the aluminum substrate, whereas Si was from the electrolyte. A small P peak was also detected (as shown in Figure S1), this phosphorus might result from the residual electrolyte. With the increase in oxidation time, the surface content of coatings had no obvious change. The EDS result was affected by the aluminum substrate, because the penetration depth of the X-ray was ~3 μm under the present conditions. Thus, the aluminum content in the thin coating formed in 5 min was unusually high.

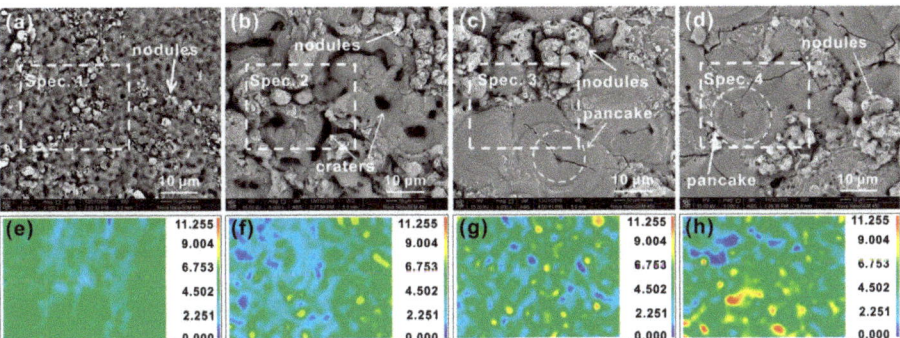

Figure 2. (**a**–**d**) Surface SEM (back-scattered electron mode) and (**e**–**h**) 3D color maps of PEO coatings formed at (**a,e**) 5 min, (**b,f**) 15 min, (**c,g**) 45 min, and (**d,h**) 60 min.

Table 1. Surface EDS analysis of PEO coatings formed at different times marked in Figure 2 (at.%).

Coating	Location	O	Al	Si	P	Si/Al	P/Al
5 min	Spec. 1	55.6	43.2	1.2	–	0.03	–
15 min	Spec. 2	62.6	30.9	5.4	1.1	0.17	0.04
45 min	Spec. 3	61.8	28.9	8.5	≤1	0.30	≤0.03
60 min	Spec. 4	62.9	29.2	7.6	≤1	0.26	≤0.03

Figure 3 shows the XRD patterns of PEO coatings prepared at different times. Both the powder sample and PEO coating sample with the substrate formed at 60 min were used for the XRD test. As shown in Figure 3d,e, the only difference between the two was that the diffraction peak of Al was present in the coating sample with the substrate. In order to reduce the effect of the aluminum substrate on the XRD results, free-standing coatings were ground to powders for the XRD tests. During the initial stages (5–15 min), the major crystalline phase in the coating was γ-Al_2O_3. The coatings formed over a longer period (45–60 min) showed a presence of γ-Al_2O_3, α-Al_2O_3, and σ-Al_2O_3.

In addition, the element distribution at the surface of the typical coating (60 min) are shown in Figure 4. As shown in Figure 4a, the surface of the PEO coating formed at 60 min was dominated by the compact pancake-like structures that were surrounded by loose nodules. EDS mappings show that Al and O were present in most regions of the coating, but the distribution of Al in the nodules was relatively low. Si and P were the main components of the nodules. EDS point analyses (Figure 4f) also revealed significant difference between the nodules and pancake-like structures.

Based on the above analysis, it can be deduced that the oxidation of aluminum and the deposition of electrolyte composition contributed to the growth of PEO coatings [28,29] in the silicate-phosphate electrolyte. The molten-shaped products resulted from the oxidation of aluminum, whereas the nodules were caused by the deposition of electrolyte compounds. Interestingly, the molten-shaped products were always surrounded by the nodules. Combined with the EDS and XRD results, the surface structure of PEO coating could be described as a molten-shaped structure of alumina surrounded by Si-rich nodules.

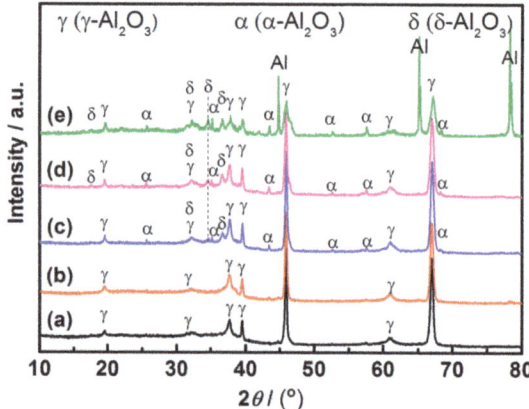

Figure 3. XRD patterns of PEO coating powder samples formed at (**a**) 5 min, (**b**) 15 min, (**c**) 45 min, and (**d**) 60 min. (**e**) XRD pattern of the PEO coating sample with a substrate formed at 60 min.

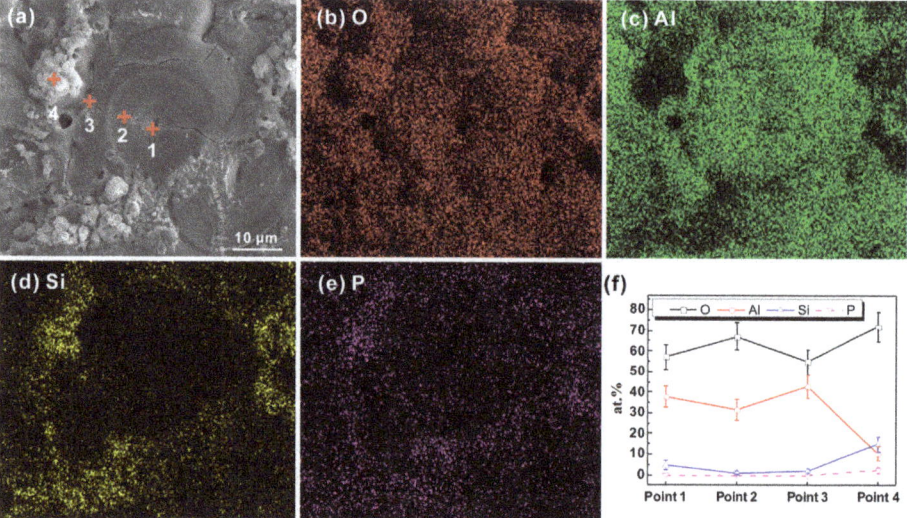

Figure 4. (**a**) Surface SEM images (secondary electron mode) of the PEO coating formed at 60 min. EDS mappings of (**b**) O, (**c**) Al, (**d**) Si, and (**e**) P. (**f**) Semi-quantitative analysis of element content at different locations marked in Figure 4a.

3.3. Internal Structure of PEO Coating

The internal structure of PEO coating was observed by exfoliating the outer layer using a single scratch tester. Figure 5 shows the plan images of the internal structure during different PEO stages. It can be seen in Figure 5 that the cohesion of the PEO coatings was poor and that spalling occurred on the edge of the scratch. The internal structure of the coatings was obvious different from the outer surface. Nodules and pancake-like structures were absent inside the coating. For the coating formed at 5 min (Figure 5a,e), the exposed internal structure was smooth. After 15 min, all the internal structure showed obvious cracks and submicron pores with the different sizes and shapes.

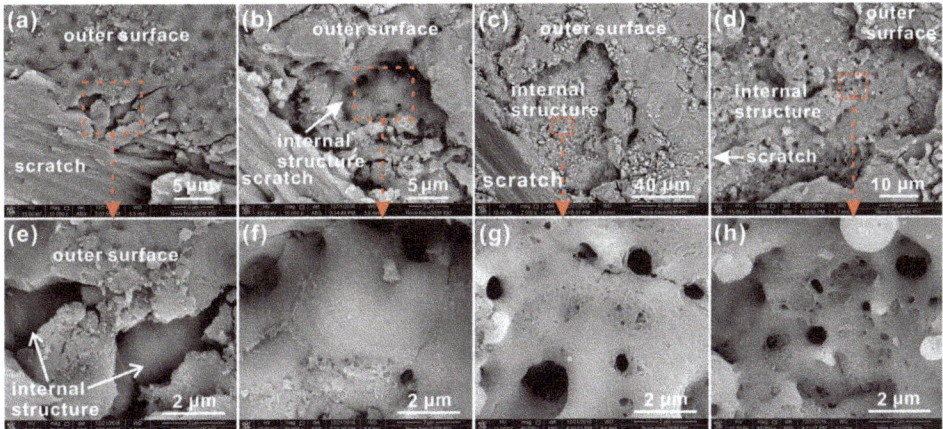

Figure 5. (**a–d**) SEM images (back-scattered electron mode) and (**e–h**) magnified SEM images of the internal structure of PEO coatings formed at (**a,e**) 5 min, (**b,f**) 15 min, (**c,g**) 45 min, and (**d,h**) 60 min.

3.4. Aluminum/Coating Interface Morphology of PEO Coating

To understand the structure evolution of the coatings during the PEO process, the aluminum/coating interface structures were obtained after the coatings were detached from the samples. Figure 6 shows the 3D and SEM images of the aluminum/coating interface at different oxidation times. The structure of the aluminum/coating interface was greatly different from the outer surface and the internal structure of the PEO coating. The aluminum/coating interface was irregular and uneven, and appeared as "hill"-like features.

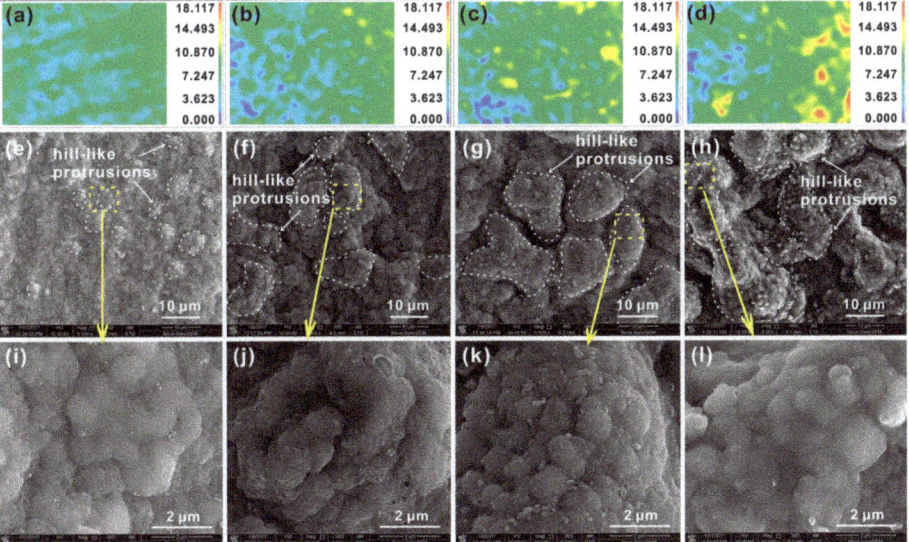

Figure 6. (**a–d**) 3D color maps, (**e–h**) SEM image, and (**i–l**) magnified SEM images of aluminum/coating interface for PEO coatings formed at (**a,e,i**) 5 min, (**b,f,j**) 15 min, (**c,g,k**) 45 min, and (**d,h,l**) 60 min.

The 3D images (Figure 6a–d) displayed that the height difference at the aluminum/coating interface raised as the oxidation time increased. This transformation indicated that the "hills" had grown with the development of the PEO process. The sites of the hill-like protrusions reflected the depression locations in the aluminum substrate. Thus, the depth of oxidation of the aluminum substrate increased with the growth of the coating.

A similarly increasing trend could be observed in the corresponding SEM images shown in Figure 6e–h. With the increase in oxidation time, the volume of the concave and convex regions at the interface increased and the boundary became smoother. According to the magnification images in Figure 6i–l, the aluminum/coating interface structure was more compact than the surface and internal structure of the PEO coating. The aluminum/coating interface was formed by a high density of cells within the cluster, and almost no pores or cracks were present. Moreover, the diameter of the cells always maintained a constant size of ~1 µm with the increase in treatment time, although the volume of the "hill"-like projections increased.

It can be inferred that the inward growth of the PEO coating was carried out such that the aluminum transformed into oxide in many valley-shaped pools as oxidation time increased. Both Figures 1 and 6 indicate that the aluminum/coating interface was a dense layer composed of many small cells within clusters embedded into the aluminum substrate.

4. Discussion

4.1. The Formation Process for 3D Structures of PEO Coating

The surface morphologies of the PEO coatings (Figure 2) demonstrate that numerous nodules surrounding the molten-shaped products were the main feature of the coating surface. The EDS results (Figures 1e,f and 4) confirmed that nodules were Si-rich products of electrolyte deposits and that the molten-shaped products were mainly oxides of the substrate. The different morphologies and components at the surface were ascribed to various kinds of discharges in the integrated discharge model [30–32]. Additionally, a ~1 µm barrier layer that consists of dense cells was present at the aluminum/coating interface. Hill-like protrusions at the aluminum/coating interface enlarged over time. According to the surface and the aluminum/coating interface, it can be deduced that molten zones were present around the plasma discharge channels due to the high temperature (~16000 ± 3500 K [33]). The molten zone was considered to be the basic unit for the formation of the coating.

Figure 7 gives a schematic diagram of the discharge at the molten zone of local coating. When the discharge occurred, the aluminum was melted and reacted with oxygen.

$$Al \rightarrow Al^{3+} + e \tag{1}$$

$$2Al^{3+} + 3O^{2-} \rightarrow Al_2O_3 \tag{2}$$

The aluminum/coating interface was an important cooling region because of the extremely high thermal conductivity (~230 W·m^{-1}·K^{-1} [34]) of the aluminum substrate. The region of the molten zone near the aluminum rapidly solidified and formed a hill-like protrusion, as shown in Figure 7a. A part of the molten products was ejected along the discharge channel to the coating surface. The coating/electrolyte interface acted as a vital cooling region, and the molten products rapidly solidified. A molten-shaped product structure formed at the coating surface.

At local high temperatures of the discharges, electrolyte will evaporate, concentrate, transform, and deposit at the coating surface to form nodules consisting of electrolyte constituents [8]. Thus, nodules rich in Si elements formed around the molten-shaped products.

In general, the following transformation process takes place.

$$H_2O + SiO_3^{2-} \rightarrow SiO_2 + 2OH^- \tag{3}$$

This analysis is confirmed by Figures 2 and 4, which show that most nodules were distributed around the molten zones and contained higher levels of Si.

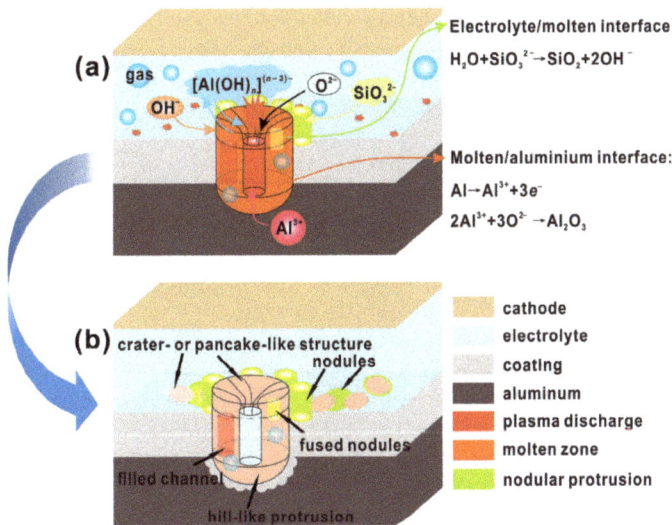

Figure 7. A schematic diagram of the discharge at the molten zone of local coating: (**a**) molten zone; (**b**) 3D structure.

Plasma discharges occurred repeatedly near the cooling region. The previously formed nodules would be broken again and incorporated into the molten products. Thus, a fresh molten-shaped product was formed after the molten zone cooling. A molten-shaped product structure of alumina surrounded by the nodules containing some electrolyte constituents was finally produced. The molten zone was generally considered to be a closed system during the cooling process. The escape of a large amount of gas was impeded, and numerous closed holes were enclosed inside the coating.

4.2. Growth Model of PEO Coating

In general, mechanisms such as "dielectric breakdown", "discharge-in-pore", and "contact glow discharge electrolysis" were the mainstream views about the plasma discharges during the PEO process [35]. Various growth models were proposed based on these mechanisms to describe the formation of the PEO coating [30–32,36]. These models illustrated the origins of plasma discharges and the relationship between discharges and coating structure. However, the aluminum/coating interface and the correlations between the growth mechanism and the 3D structure were not considered in these models. Many studies have shown that the discharges become more powerful and extend as the coating thickens [30,37]. Therefore, the molten zones caused by discharges will also enlarge and last longer as the coating thickens.

According to the surface and aluminum/coating interface morphologies (Figures 2 and 6), the molten-shaped products, the nodules at the surface, and the hill-like structures at the aluminum/coating interface tended to decrease in number and increase in size as the coating thickened. Since the molten zones were the main routes by which a new coating was formed at local regions, continuous changes and overlaps of the molten zones would lead to the evolution of the coating structure. Based on the above results, a growth model is here proposed to explain the correlations between the molten zones and 3D structure evolution during the PEO process.

Figure 8 provides the growth model of the PEO coating. The formation of a dielectric film on the surface of the sample was a necessary condition for the plasma discharge. The dielectric film would be damaged once plasma discharge occurred.

In the early stage (5 min), as the discharges were weak, the previously formed anodic film had not been significantly damaged. The major crystalline phase in the coating was γ-Al_2O_3. The height difference of the aluminum/coating interface was lower, as demonstrated by the weak hill-like features. At this stage, obvious porosity defects were difficult to find, because the gas could easily escape from the molten zone, as shown in Figure 8a.

With the increase in oxidation time, a thin and dense barrier layer near the coating/substrate interface was clearly visible in the coating formed at 15 min (Figures 1b, 6f, and 7j). The higher height difference of the aluminum/coating interface (Figure 6b) suggested an increased depth of the oxidation of the aluminum substrate. The larger nodules at the surface (Figure 2b) indicated an increased area and a longer duration of the molten zone, which was caused by a stronger discharge. Additionally, as shown in Figures 1b and 2b, trench-like open pores were present in the center of the molten-shaped products, which was very common and usually appeared in the thinner PEO coatings on Al [38], Mg [31], and Ti [6]. A reasonable explanation for the open pores was that the amount of molten products was insufficient to complement the plasma discharge channels in the cooling process. Now, the model of the PEO coating and discharge was shown in Figure 8b. At this stage, the molten-shaped product presented a "crater"-like morphology with an open pore in the center.

As the oxidation time increased, the dielectric breakdown of the thicker coating became difficult, and the discharge events appeared to be more powerful. Longer cooling periods might occur in this situation, and there is a strong tendency for discharges in the cascade [33]. It was reasonable to consider that the long-lasting molten zones were easier to form in the later stages. As oxidation time increased, the coating tended to form more high temperature phases of α-Al_2O_3 and σ-Al_2O_3. Consequently, the larger pancake-like structures with closed center pores formed on the coating over 45 min, and the hill-like protrusions at the aluminum/coating interface were further enlarged. Figure 8c,d illustrates the model of these coatings.

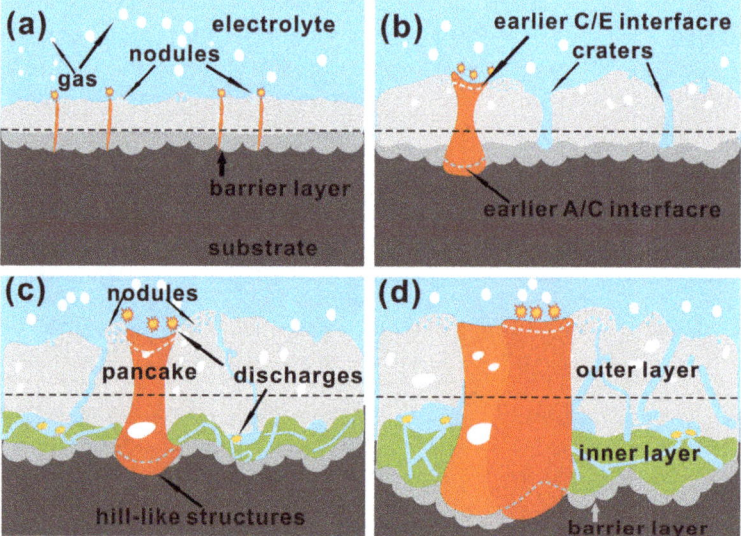

Figure 8. A growth and 3D structure model of the PEO coating at different stages: (**a**) breakdown of dielectric film under plasma discharges; (**b**) formation of PEO coating with open pores; (**c**) initial formation of three-layer structure; and (**d**) further evolution of three-layer structure.

It has been proven that excessive gas is released during discharge [39] and that gas might be generated near the substrate/coating interface [9,15]. In this work, cracks or pores would form when high-pressure gases escaped from the molten zone. Many gas bubbles could not escape in time, which left the closed, spherical pores outside the coating. The volume shrinkage during the solidification of the molten regions and the gas expansion left large cavities near the substrate/coating interface. Different discharges occurred repeatedly at adjacent locations, which caused an overlap of molten zones, implying in turn that channels, cracks, and large cavities formed in the coating. Thus, a fine, interconnected porosity network structure was formed in the PEO coating. The porosity network caused the electrolyte to penetrate into the large cavities, and a secondary coating/electrolyte interface formed inside the coating. Further discharges were likely to initiate at the base of the cavities, so a thinner, finely porous inner-layer coating was formed, as shown in Figure 2c,d. As the coating grew, the thickness of the inner-layer continuously increased. It can be inferred that a series of reactions (evaporation, dehydration, and deposition) occurred in the large cavities. Thus, a higher proportion of electrolyte species was present at the edges of large cavities (Figure 1e,f).

5. Conclusions

The 3D structure of a PEO coating on aluminium, including the surface, the internal structure, the aluminum/coating interface, and the fracture cross-section morphology, was obtained. The PEO coating surface can be described as molten-shaped products of alumina surrounded by Si-rich nodules. A barrier layer consisting of clustered cells was present at the aluminum/coating interface. The PEO coating gradually evolved into a distinct three-layer structure as the coating thickened, including a barrier layer, an inner layer with enclosed pores, and an outer layer with a rough surface. Interestingly, obvious cavities appeared between the inner and outer layers in the thicker coatings.

PEO coatings were grown via the oxidation of aluminum and the deposition of electrolyte compounds. The oxidation of aluminum resulted in a crater- or pancake-like molten-shaped product, whereas the deposition of electrolyte compounds usually formed nodules at the coating surface.

During the PEO process, molten zones were formed around the plasma discharges. The thickening of the coating mainly depended on the forming, closing, and repeated movement of molten zones. The uneven cooling rates around the molten zones resulted in a distinction between the coating surface and the aluminum/coating interface structure. At different discharge periods, the intensity and duration of discharges determined the volume and lifetime of the molten zones, which resulted in various 3D microstructures of the PEO coating.

Supplementary Materials: The following are available online at http://www.mdpi.com/2079-6412/8/3/105/s1, Figure S1: the EDS spectrum of PEO coating formed at 60 min.

Acknowledgments: This work was supported by the National Natural Science Foundation of China (Grant No. 51361025) and the Natural Science Foundation of Jiangxi Province (Grant No. 20171BAB216006).

Author Contributions: Xiaohui Liu, Shuaixing Wang, and Xinyi Li performed the PEO treatments; Nan Du and Qing Zhao contributed reagents, materials, and analysis tools; Xiaohui Liu and Shuaixing Wang characterized the PEO coatings and wrote the paper.

Conflicts of Interest: The authors declare no conflict of interest.

References

1. Hussein, R.O.; Northwood, D.O.; Nie, X. Coating growth behavior during the plasma electrolytic oxidation process. *J. Vac. Sci. Technol. A* **2010**, *28*, 766–773. [CrossRef]
2. Sharifi, H.; Aliofkhazraei, M.; Darband, G.; Rouhaghdam, A.S. Characterization of PEO nanocomposite coatings on titanium formed in electrolyte containing atenolol. *Surf. Coat. Technol.* **2016**, *304*, 438–449. [CrossRef]
3. Wang, S.; Zhao, Q.; Liu, D.; Du, N. Microstructure and elevated temperature tribological behavior of TiO_2/Al_2O_3 composite ceramic coating formed by microarc oxidation of Ti6Al4V alloy. *Surf. Coat. Technol.* **2015**, *272*, 343–349. [CrossRef]

4. Gowtham, S.; Hariprasad, S.; Arunnellaiappan, T.; Rameshbabu, N. An investigation on ZrO_2 nano-particle incorporation, surface properties and electrochemical corrosion behaviour of PEO coating formed on Cp-Ti. *Surf. Coat. Technol.* **2017**, *313*, 263–273.
5. Yang, J.; Lu, X.; Blawert, C.; Di, S.; Zheludkevich, M.L. Microstructure and corrosion behavior of Ca/P coatings prepared on magnesium by plasma electrolytic oxidation. *Surf. Coat. Technol.* **2017**, *319*, 359–369. [CrossRef]
6. Zhang, X.; Aliasghari, S.; Němcová, A.; Burnett, T.; Kuběna, I.; Šmíd, M.; Thompson, G.; Skeldon, P.; Withers, P. X-ray computed tomographic investigation of the porosity and morphology of plasma electrolytic oxidation coatings. *ACS Appl. Mater. Interfaces* **2016**, *8*, 8801–8810. [CrossRef] [PubMed]
7. Li, Q.; Liang, J.; Liu, B.; Peng, Z.; Wang, Q. Effects of cathodic voltages on structure and wear resistance of plasma electrolytic oxidation coatings formed on aluminium alloy. *Appl. Surf. Sci.* **2014**, *297*, 176–181. [CrossRef]
8. Matykina, E.; Arrabal, R.; Skeldon, P.; Thompson, G.E. Incorporation of zirconia nanoparticles into coatings formed on aluminium by AC plasma electrolytic oxidation. *J. Appl. Electrochem.* **2008**, *38*, 1375–1383. [CrossRef]
9. Tillous, K.; Toll-Duchanoy, T.; Bauer-Grosse, E.; Hericher, L.; Geandier, G. Microstructure and phase composition of microarc oxidation surface layers formed on aluminium and its alloys 2214-T6 and 7050-T74. *Surf. Coat. Technol.* **2009**, *203*, 2969–2973. [CrossRef]
10. Curran, J.A.; Clyne, T.W. Porosity in plasma electrolytic oxide coatings. *Acta Mater.* **2006**, *54*, 1985–1993. [CrossRef]
11. Curran, J.A.; Clyne, T.W. Thermo-physical properties of plasma electrolytic oxide coatings on aluminium. *Surf. Coat. Technol.* **2005**, *199*, 168–176. [CrossRef]
12. Kasalica, B.; Radić-Perić, J.; Perić, M.; Petković-Benazzouz, M.; Belča, I.; Sarvan, M. Themechanism of evolution ofmicrodischarges at the beginning of the PEO process on aluminum. *Surf. Coat. Technol.* **2016**, *298*, 24–32. [CrossRef]
13. Liu, C.; He, D.; Yan, Q.; Huang, Z.; Liu, P.; Li, D.; Jiang, G.; Ma, H.; Nash, P.; Shen, D. An investigation of the coating/substrate interface of plasma electrolytic oxidation coated aluminum. *Surf. Coat. Technol.* **2015**, *280*, 86–91. [CrossRef]
14. Liu, C.; Liu, P.; Huang, Z.; Yan, Q.; Guo, R.; Li, D.; Jiang, G.; Shen, D. The correlation between the coating structure and the corrosion behavior of the plasma electrolytic oxidation coating on aluminum. *Surf. Coat. Technol.* **2016**, *286*, 223–230. [CrossRef]
15. Tillous, E.K.; Toll-Duchanoy, T.; Bauer-Grosse, E. Microstructure and 3D microtomographic characterization of porosity of MAO surface layers formed on aluminium and 2214-T6 alloy. *Surf. Coat. Technol.* **2009**, *203*, 1850–1855. [CrossRef]
16. Moon, S.; Arrabal, R.; Matykina, E. 3-Dimensional structures of open-pores in PEO films on AZ31 Mg alloy. *Mater. Lett.* **2015**, *161*, 439–441. [CrossRef]
17. Yerokhin, A.L.; Lyubimov, V.V.; Ashitkov, R.V. Phase formation in ceramic coatings during plasma electrolytic oxidation of aluminium alloys. *Ceram. Int.* **1998**, *24*, 1–6. [CrossRef]
18. Xue, W.; Deng, Z.; Chen, R.; Zhang, T. Growth regularity of ceramic coatings formed by microarc oxidation on Al–Cu–Mg alloy. *Thin Solid Films* **2000**, *372*, 114–117. [CrossRef]
19. Liu, X.; Li, G.; Xia, Y. Investigation of the discharge mechanism of plasma electrolytic oxidation using Ti tracer. *Surf. Coat. Technol.* **2012**, *206*, 4462–4465. [CrossRef]
20. Sundararajan, G.; Krishna, L.R. Mechanisms underlying the formation of thick alumina coatings through the MAO coating technology. *Surf. Coat. Technol.* **2003**, *167*, 269–277. [CrossRef]
21. Li, J.; Cai, H.; Xue, X.; Jiang, B. The outward–inward growth behavior of microarc oxidation coatings in phosphate and silicate solution. *Mater. Lett.* **2010**, *64*, 2102–2104. [CrossRef]
22. Matykina, E.; Arrabal, R.; Scurr, D.J.; Baron, A.; Skeldon, P.; Thompson, G.E. Investigation of the mechanism of plasma electrolytic oxidation of aluminium using ^{18}O tracer. *Corros. Sci.* **2010**, *52*, 1070–1076. [CrossRef]
23. Gu, W.; Lv, G.; Chen, H.; Chen, G.; Feng, W.; Yang, S. Characterisation of ceramic coatings produced by plasma electrolytic oxidation of aluminum alloy. *Mater. Sci. Eng. A* **2007**, *447*, 158–162. [CrossRef]
24. Li, W.; Qian, Z.; Liu, X.; Zhu, L.; Liu, H. Investigation of micro-arc oxidation coating growth patterns of aluminum alloy by two-step oxidation method. *Appl. Surf. Sci.* **2015**, *356*, 581–586. [CrossRef]

25. Zhu, L.; Guo, Z.; Zhang, Y.; Li, Z.; Sui, M. A mechanism for the growth of a plasma electrolytic oxide coating on Al. *Electrochim. Acta* **2016**, *208*, 296–303. [CrossRef]
26. Li, Q.; Yang, W.; Liu, C.; Wang, D.; Liang, J. Correlations between the growth mechanism and properties of micro-arc oxidation coatings on titanium alloy: Effects of electrolytes. *Surf. Coat. Technol.* **2017**, *316*, 162–170. [CrossRef]
27. Monfort, F.; Berkani, A.; Matykina, E.; Skeldon, P.; Thompson, G.E.; Habazaki, H.; Shimizu, K. A tracer study of oxide growth during spark anodizing of aluminum. *J. Electrochem. Soc.* **2005**, *152*, C382–C387. [CrossRef]
28. Cheng, Y.; Wang, T.; Li, S.; Cheng, Y.; Cao, J.; Xie, H. The effects of anion deposition and negative pulse on the behaviours of plasma electrolytic oxidation (PEO)—A systematic study of the PEO of a Zirlo alloy in aluminate electrolytes. *Electrochim. Acta* **2017**, *225*, 47–68. [CrossRef]
29. Li, Q.; Liu, C.; Yang, W.; Liang, J. Growth mechanism and adhesion of PEO coatings on 2024Al alloy. *Surf. Eng.* **2017**, *33*, 760–766. [CrossRef]
30. Hussein, R.O.; Nie, X.; Northwood, D.O.; Yerokhin, A.; Matthews, A. Spectroscopic study of electrolytic plasma and discharging behaviour during the plasma electrolytic oxidation (PEO) process. *J. Phys. D-Appl. Phys.* **2010**, *43*, 105203–105215. [CrossRef]
31. Hussein, R.O.; Nie, X.; Northwood, D.O. An investigation of ceramic coating growth mechanisms in plasma electrolytic oxidation (PEO) processing. *Electrochim. Acta* **2013**, *112*, 111–119. [CrossRef]
32. Cheng, Y.L.; Xue, Z.G.; Wang, Q.; Wu, X.Q.; Matykina, E.; Skeldon, P.; Thompson, G.E. New findings on properties of plasma electrolytic oxidation coatings from study of an Al–Cu–Li alloy. *Electrochim. Acta* **2013**, *107*, 358–378. [CrossRef]
33. Dunleavy, C.S.; Golosnoy, I.O.; Curran, J.A.; Clyne, T.W. Characterisation of discharge events during plasma electrolytic oxidation. *Surf. Coat. Technol.* **2009**, *203*, 3410–3419. [CrossRef]
34. Lee, S.; Kwon, S.; Lee, J.-C.; Lee, S.-W. Thermophysical properties of aluminum 1060 fabricated by equal channel angular pressing. *Int. J. Thermophys.* **2012**, *33*, 540–551. [CrossRef]
35. Yerokhin, A.L.; Snizhko, L.O.; Gurevina, N.L.; Leyland, A.; Pilkington, A.; Matthews, A. Discharge characterization in plasma electrolytic oxidation of aluminium. *J. Phys. D-Appl. Phys.* **2003**, *36*, 2110–2120. [CrossRef]
36. Sobrinho, P.H.; Savguira, Y.; Ni, Q.; Thorpe, S.J. Statistical analysis of the voltage-time response produced during PEO coating of AZ31B magnesium alloy. *Surf. Coat. Technol.* **2017**, *315*, 530–545. [CrossRef]
37. Nominé, A.; Troughton, S.C.; Nominé, A.V.; Henrion, G.; Clyne, T.W. High speed video evidence for localised discharge cascades during plasma electrolytic oxidation. *Surf. Coat. Technol.* **2015**, *269*, 125–130. [CrossRef]
38. Dehnavi, V.; Luan, B.L.; Liu, X.Y.; Shoesmith, D.W.; Rohani, S. Correlation between plasma electrolytic oxidation treatment stages and coating microstructure on aluminum under unipolar pulsed DC mode. *Surf. Coat. Technol.* **2015**, *269*, 91–99. [CrossRef]
39. Snizhko, L.O.; Yerokhin, A.L.; Gurevina, N.L.; Patalakha, V.A.; Matthews, A. Excessive oxygen evolution during plasma electrolytic oxidation of aluminium. *Thin Solid Films* **2007**, *516*, 460–464. [CrossRef]

 © 2018 by the authors. Licensee MDPI, Basel, Switzerland. This article is an open access article distributed under the terms and conditions of the Creative Commons Attribution (CC BY) license (http://creativecommons.org/licenses/by/4.0/).

Article

Particle Characterisation and Depletion of Li₂CO₃ Inhibitor in a Polyurethane Coating

Anthony Hughes [1,2,*], James Laird [1], Chris Ryan [1], Peter Visser [3,4], Herman Terryn [3,5] and Arjan Mol [3]

1. CSIRO Minerals Resources Flagship, Clayton, Victoria 3169, Australia; Jamie.Laird@csiro.au (J.L.); chris.ryan@csiro.au (C.R.)
2. Institute for Frontier Materials, Deakin University, Waurn Ponds, Geelong, Victoria 3216, Australia
3. Department of Materials Science and Engineering, Delft University of Technology, Mekelweg 2, 2628 CD Delft, The Netherlands; P.Visser-1@tudelft.nl (P.V.); Herman.Terryn@vub.be (H.T.); J.M.C.Mol@tudelft.nl (A.M.)
4. AkzoNobel Specialty Coatings, Rijksstraatweg 31, 2171 AJ Sassenheim, The Netherlands
5. Department of Materials and Chemistry, Research Group Electrochemical and Surface Engineering, Vrije Universiteit Brussel, Pleinlaan 2, 1050 Brussels, Belgium
* Correspondence: Tony.Hughes@csiro.au; Tel.: +61-3-9545-2705; Fax: +61-3-9544-1128

Received: 30 May 2017; Accepted: 14 July 2017; Published: 21 July 2017

Abstract: The distribution and chemical composition of inorganic components of a corrosion-inhibiting primer based on polyurethane is determined using a range of characterisation techniques. The primer consists of a Li_2CO_3 inhibitor phase, along with other inorganic phases including TiO_2, $BaSO_4$ and Mg-(hydr)oxide. The characterisation techniques included particle induced X-ray and γ-ray emission spectroscopies (PIXE and PIGE, respectively) on a nuclear microprobe, as well as SEM/EDS hyperspectral mapping. Of the techniques used, only PIGE was able to directly map the Li distribution, although the distribution of Li_2CO_3 particles could be inferred from SEM through using backscatter contrast and EDS. Characterisation was also performed on a primer coating that had undergone leaching in a neutral salt spray test for 500 h. Overall, it was found that Li_2CO_3 leaching resulted in a uniform depletion zone near the surface, but also much deeper local depletion, which is thought to be due to the dissolution of clusters of Li_2CO_3 particles that were connected to the external surface/electrolyte interface.

Keywords: primer; Li-inhibited; AA2024; polyurethane; SEM; EDS; PIXE; PIGE; leaching; pigments

1. Introduction

Over recent decades, there has been a widespread search for alternatives to chromate inhibitors in paints for many applications where corrosion is a threat to aesthetic and structural quality. Many new inhibitor systems have been the subject of extensive research, including rare earths [1–14], vanadates [15,16], organic compounds [17–23], sacrificial particles or functional properties in coatings [24–28], double-layered hydroxides containing inhibitors [29–41] and, in many cases, combinations of these. The search for alternatives is probably most intense for aerospace applications, where chromate inhibitors have been the mainstay of corrosion prevention for many decades. This is because chromate has a proven performance, particularly in the parts of aerostructures that are difficult to access, where many years may pass between inspections. It is in these applications that chromate has demonstrated its reliability [42].

Recently, Visser et al. reported on the promising potential of Li-based inhibitors as chromate replacements for application to aluminium alloys used in the aerospace industry [43]. The successful inhibition of Al, exposed at defects through a primer by Li inhibitors, was proposed to be due to the formation of a hydrated aluminium oxide incorporating Li [44–46]. The source of the Li was from Li_2CO_3 particles added as a leachable inhibitor to the paint system. The successful application of Li-based inhibitors in coatings, however, requires a detailed knowledge of how to incorporate the inhibitor into a primer formulation, as well as an understanding of the mechanism of leaching. Thus, detailed characterisation of the coating in the as-formulated state, as well as after exposure to conditions where inhibitor leaching occurs, is required.

The objective of this work is to characterise a polyurethane primer, particularly its inorganic components (Li_2CO_3 inhibitor, TiO_2, $BaSO_4$ and Mg-(hydr)oxide) in the as-formulated state, as well as after neutral salt spray (NSS) exposure. Comparing the primer without NSS exposure to that after NSS exposure will provide important insights into changes in the primer chemistry, including changes to the inorganic components, particularly the Li_2CO_3 inhibitor, resulting from NSS exposure. From a characterisation perspective, Li is one of the more difficult elements to detect in the periodic table. This is due both to the small number of electrons as well as the low interaction cross section for techniques based on electron or photon interaction. On the other hand, the Li nucleus has a relatively high cross section for proton interaction (depending on the proton energy) resulting in γ-ray emission. The method that utilises this interaction is called particle-induced γ-ray emission (PIGE), and a preceding paper on the application of this technique on the Li_2CO_3-loaded primer studied here has recently been reported [47]. Both PIGE and X-ray emission (PIXE) occur using proton beams, with the X-ray emission resulting from the same transitions as can be seen using energy-dispersive X-ray spectroscopy, but with proton excitation rather than electron excitation. The use of these techniques is not widespread in the corrosion and coatings communities, since nuclear microprobes using MeV ion beams from particle accelerators are rare compared to standard instruments available to research and industry. However, PIXE has been used for corrosion studies [48,49], as well as studies into inhibitor depletion at defects in coatings [50]. In this study, PIGE, supported by PIXE and SEM/EDS has been used to characterise the inorganic components of the primer system, including the Li_2CO_3, Mg-(hydr)oxides, $BaSO_4$ and TiO_2 particles. Of course, after NSS, the focus is on the change in the Li distribution, since this is the leachable component of the system.

2. Experimental

2.1. Materials and Sample Preparation

The primer (coating) was a high-solids formulation based on a polyurethane resin with polyisocynate crosslinker, and formulated to a pigment volume concentration (PVC) of 30%, as described elsewhere [43]. The inorganic pigments included Li_2CO_3, Mg-(hydr)oxide and $BaSO_4$ fillers and TiO_2. Trace element analyses of the various inorganic components of the paint indicated that the $BaSO_4$ contained 0.9%m/m Sr, and small amounts of Si, Ca, Al and Ti. The MgO contained 1500–1900 ppm by weight of Ca, 400–700 ppm by weight K and lesser amounts of other elements (Table 1). The Li_2CO_3 contained alkali metals (Na, K) in the range 400–800 ppm by weight. Particle size distributions for these additives were determined by dispersing in a solvent, which was methylethylketone for Mg-(hydr)oxide, TiO_2 and Li_2CO_3, where water was used for the $BaSO_4$. The $BaSO_4$ particles had the largest particles (up to 50 μm) and the largest spread in particles size. The TiO_2 particles were the smallest (up to 14 μm) and slightly smaller than the Mg-(hydr)oxide. The Li_2CO_3 had the largest size at the lowest end of the distribution and ranged up to 18 μm.

AA2024-T3 was used as a substrate for coating; typical breakdown for this alloy is reported elsewhere [51]. The AA2024-T3 was prepared by standard anodising according to aerospace requirements (AIPI 02-01-003) at Premium AEROTEC, Bremen Germany. This included the following steps: degrease, alkaline clean, acid desmutting followed by anodising in tartaric-sulphuric acid to produce a 2–3 μm thick oxide layer. Subsequently, the primer was applied by spraying using a high-volume, low-pressure (HVLP) spray gun in a single pass to achieve a dry film thickness of approximately 30 μm. In practice, the coating was typically 30 to 40 μm. Finally, the primer was cured for 16 h at 23 °C/55% RH, followed by a 30 min baking cycle at 80 °C.

Table 1. Chemical composition of inorganic additives used in this study.

Element	Li$_2$CO$_3$ (mg/kg)	TiO$_2$ (%m/m)	MgO (mg/kg)	BaSO$_4$ (%m/m)
Al	5	1.4	70–77	0.1
Ca	91–98	–	1500–1900	0.1
Na	660–810	–	180–210	0.4
Ba	2	–	2–5	res
Si	–	–	–	0.4
Sr	–	–	–	0.9
Ti	–	–	–	0.1
Zr	–	0.4	–	–
Fe	2	–	73–78	–
Mn	–	–	13–15	–
Ni	–	–	5–6	–
K	400–700	–	200–300	–
Mg	39–40	–	–	1

2.2. NSS Exposure

The primer was exposed to neutral salt spray (NSS) for 500 h in a test chamber operated according to ASTM B117 [52].

2.3. Particle Induced Y (PIGE) and X-ray Emission (PIXE)

PIXE and PIGE were performed on the CSIRO beamline attached to a pelletron at the University of Melbourne [53]. Three MeV protons were focussed at the target plane to around 2 μm using a separated quintuplet lens designed for optimal balance between high spatial resolution and maximum beam current. For this work, beam currents were typically in the 0.5–1.0 nA range. A large area Ge(Li) γ-ray detector was placed approximately 5 mm directly behind the sample for a maximum acceptance solid angle. A LiF crystal and pure Al were used for the calibration of the γ-ray detector energy axis. For PIXE, the 100 mm^2 Ge(Li) detector was mounted at 45° to the incident proton beam, and around 3 to 4 mm from the sample. A 100 μm thick pure Al filter was placed in front of the detector to accommodate trace level heavy element detection limits. Scan areas chosen for analysis varied, but generally ranged from 10 to 50 μm × 200 μm. The analysis depth was approximately 10–20 μm for both methods. A schematic for data collection using MicroDAQ [54] is presented in Figure 1. For data collection, the sample is moved in a grid of points under the proton beam. At each point, PIGE and PIXE spectra were collected, forming a pixel in a map and making a hyperspectral data set.

After collection, further data analysis was performed using GeoPIXE [55] where regions of interest (ROI) such as the primer, aluminium alloy or depletion zones were examined in more detail by extracting spectra from each of these ROIs. Both the Li 429 and 478 keV lines were considered for PIGE Li analysis, but only the 429 keV peak was employed due to its greater surface sensitivity [47]. In this case, inelastic proton scattering from the nucleus (written ^{7}Li(p, p', γ)) generates a clear 429 keV γ-ray signature, making PIGE an excellent technique for following changes in the Li distribution, since Li comes only from the primer in this study [56]. For PIXE, the signature K and L-series X-ray emission lines were used for element identification.

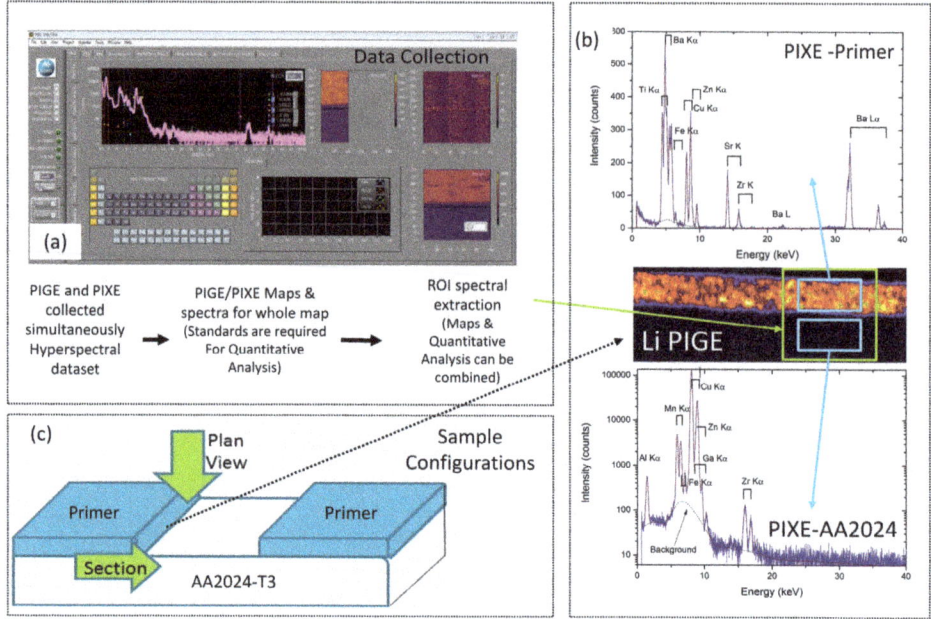

Figure 1. (**a**) Flowchart and schematic representation of the collection and analysis of X-ray emission (PIXE) and γ-ray emission (PIGE) data. The sample is scanned under the beam, and at each step a pixel is generated containing a PIGE and PIXE spectrum, thus forming a hyperspectral dataset; (**b**) After the end of the data collection, a total spectrum for the sample must be analysed to extract the regions of interest (ROI), from which quantitative analysis can be performed if standards have been collected. Examples of maps, ROI in plan view and PIGE and PIXE spectra from the green region within the primer; (**c**) Samples can be examined in section or plan view, depending on how they are mounted.

2.4. Scanning Electron Microscopy

Scanning electron microscopy was performed on a FEI Quanta 400 field emission, environmental SEM (ESEM) under high vacuum conditions. The samples were examined in a number of configurations. All samples were coated with around 200 Å of carbon. Samples examined in sections were first cut using sheet metal cutters, mounted in conducting bakerlite ground, and then polished. Grinding started with SiC papers (320 through to 2400 SiC), followed by polishing on diamond (8, 3, 1, 0.25 μm) under a non-aqueous medium. Secondary electron and backscattered electron imaging were performed using beam energies of 10 kV, and probe currents of approximately 140–145 pA. Some additional imaging was performed at higher energies to obtain information on subsurface particles, as described in the Results section. Quantmaps were generated using standardless approaches; however, the quantmaps were only used to separate the overlapping signals of the Ti K-series lines from the Ba L-series lines by curve fitting rather than for quantitative analysis.

3. Results

The results section of this paper is divided into the characterisation of the inorganic components in a polyurethane coating without exposure to NSS and after 500 h NSS exposure. Of particular interest is the Li distribution in the coating, since this is the active inhibitor component that is expected to leach out of the coating during NSS exposure.

3.1. Characterisation of the Primer Prior to Leaching

3.1.1. SEM/EDS

Figure 2 gives an overview of a section through the primer. In this section, the primer had a thickness in the vicinity of 35 ± 5 μm, contained a high level of solids and was applied to an anodised layer that was around 2–3 μm, as described in the Experimental section. The primer itself had a high level of inorganics (PVC was approximately 30%), which is reflected in the high density of particles in Figure 2. The brightest particles in the backscatter image are $BaSO_4$, which are the easiest inorganic components to identify. They are generally angular with a range of sizes (slightly less than 1 μm to over 10 μm, which is consistent with around 90% of the particle sizes for this additive (Table 2)), and an aspect ratio slightly larger than one (Figure 3b). There were another group of particles with very little contrast difference from the polyurethane containing Mg, which were assumed to be a mixture of Magnesium oxides and hydroxides, and will be referred to as Mg-(hydr)oxide in the rest of the paper. In many instances, they appeared to have a layered structure where the layers had a thickness typically 250 nm and lengths with a minimum size of around 1 μm, and typically 5–10 μm (Figure 3c), which was again consistent with the particle size distribution determined from the dispersed particles (Table 2). Mg-(hydr)oxide particles without this structure were assumed to be rotated so that the layers were viewed from the top (Figure 3a). The TiO_2 was not easily distinguished on the basis of backscatter contrast, as it was similar to the smaller particles of $BaSO_4$. Finally, there are dark particles (indicated in Figure 3a) in the film that show C and O peaks, but no significant levels of Ba (from $BaSO_4$), Mg (from Mg(hydr)oxide) or Ti (from TiO_2), implying that they are probably the Li_2CO_3 particles (Li cannot be detected in standard EDS). The sizes of these particles were similar to those of the free particle size distribution for the Li_2CO_3 particles (Table 2). This last category of particles has similar greyscale contrast to voids in the coating, making it difficult to distinguish the two without closer examination.

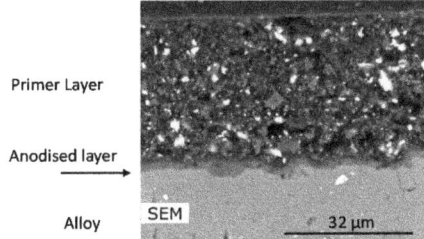

Figure 2. Backscattered electron images of sections of the primer prior to exposure to NSS. The mounting medium is at the top of the image, the primer is in the centre, and the AA2024-T3 is at the bottom of all images.

Table 2. Inorganic additive size distribution presented as the percentage of particles at a particular size in microns.

Additive	10%	50%	90%	99%
MgO	3.91	8.48	14.58	20.23
Li_2CO_3	4.78	8.87	13.44	18.26
TiO_2	3.22	5.64	8.84	13.94
$BaSO_4$	3.82	6.48	10.19	49.68

Positive identification of each of the inorganic phases using EDS alone is not straightforward. The inorganic particles sizes ranged from less than a micron up to 10 μm for larger particles, which meant that only the large particles could be sampled using EDS, with some certainty that interaction volume effects had been minimised. This can be seen in their respective spectra, where each type of particle typically contains some signal from other particles due to the interaction volume effect (Figure 4). This effect is largest for the smallest particles, which are the TiO_2 particles. Compositions (expressed as ratios of major elements) for the larger $BaSO_4$ and Mg-(hydr)oxide particles are presented in Table 3. The analyses indicate for $BaSO_4$ that the composition is close to stoichiometric, with perhaps a small underestimation of O. For Mg-(hydroxyl)oxide, the data indicates a mixture of MgO and $Mg(OH)_2$. Only C and O were detected in any significant amount for the particles thought to be Li_2CO_3, but, given that the samples were carbon-coated prior to analysis, it was not possible to conclude anything definitive from the quantitative analyses of these particles. It was not possible to determine the composition of the TiO_2 particles because of their small size (Figure 3d). This was not just due to the sampling volume containing some of the polymer matrix, but it might also contain other subsurface inorganic particles (see Appendix A).

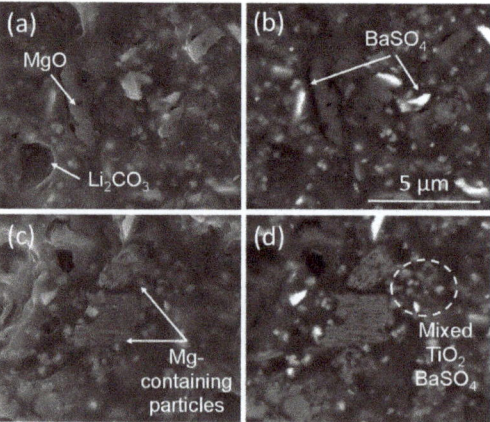

Figure 3. Examples of inorganic particles that make up the sample configurations for measuring Li depletion. Examples of (**a**) MgO and Li_2CO_3 parttticles, (**b**) $BaSO_4$ particles, (**c**) Mg-containing particles and (**d**) mixed TiO_2 and $BaSO_4$ particles. The dark areas spots show where point analyses have been performed.

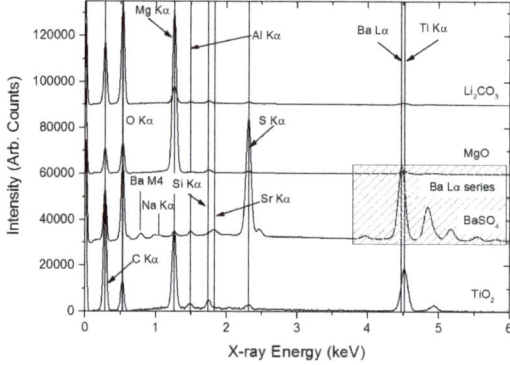

Figure 4. Typical X-ray spectra for the different inorganic components of the primer system. The arrows either point to that indicated the elements positions or sit above them. It can be seen that for each compound, there is some level of "contamination" due to sampling volume effects or small particles. This is most evident for the small TiO_2 particles that show significant Mg levels as well as S, Ba and a little Al.

Table 3. Composition (at %) from EDS analysis of inorganic particles in the primer. Each datum is an average of five determination on large particles. (Data overaged over four analyses).

Particle Type	No NSS	500 h NSS
$BaSO_4$ (Ba:S:O)	1.00:1.04 ± 0.05:3.89 ± 0.71	1.00:0.98 ± 0.01:3.51 ± 0.26
Mg-(hydr)oxide (Mg:O)	1.00:1.64 ± 0.11	1.00:0.93 ± 0.15
Li_2CO_3 (O:C)	1.72 ± 0.21	2.02 ± 0.32

Therefore, as discussed in the appendix, quantitative mapping derived from standardless fitting of the EDS spectra from hyperspectral data was used to generate elemental maps (Figure 5). The backscatter electron contrast shows several different types of particles in the primer cross section in Figure 5a, and the phases are identified in Figure 5b, which is a four-colour map of O (red) Mg (blue), Ba (green) and S (yellow). Figure 5c shows the Ti-containing particles (pink), the $BaSO_4$ particles, and highlights the Mg-(hydr)oxide particles. In both Figure 5c,d, there are particles containing O, but none of Ti, Mg or Ba; these particles are attributed to Li_2CO_3. Figure 5a–d all show an oxide at the interface, which is the anodised coating. S was detected in this layer, presumably due to the incorporation of SO_4^{2-} ions from the anodising process (Figure 5c) [57].

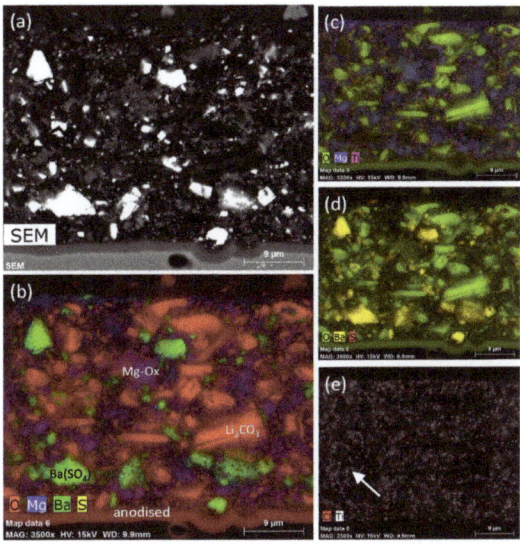

Figure 5. EDS mapping results for sample without exposure to neutral salt spray (NSS). (**a**) Backscattered electron image and maps derived from quantitative analysis, and composite maps for (**b**) O-Mg-Ba-S with phase labelling (**c**) O-Mg-Ti, (**d**) O-Ba-S and (**e**) Cl-Ti map. (N.B. colour mixing is not the same as three-colour mixing). The Li_2CO_3 was assigned on the basis that only O and C were detected at any significant levels. The arrow tip in (**e**) indicates where a very low amount of chlorine was detected.

From Figure 5c, it can be seen that there was a homogeneous distribution of Mg-(hydr)oxide particles in the coating, with larger particles appearing to be randomly distributed throughout the coating. The smaller Mg-(hydr)oxide particles also appear homogenously distributed within the coating. Similarly, Figure 5c suggests a homogeneous distribution of TiO_2 particles. The large $BaSO_4$ particles tend to be present as small clusters of two or three particles, which are randomly distributed throughout the coating, whereas the smaller $BaSO_4$ particles appear more evenly distributed. Finally, Figure 5e is a map showing the Ti and Cl distribution. There is only one region where a very small

Cl signal was detected (in the vicinity of the tip of the white arrow in Figure 5e at the periphery of a BaSO$_4$ particle). The rest of the contrast is due to the presence of Ti. This image is included for later comparison with the samples that had undergone 500 h exposure to NSS, and is discussed later.

3.1.2. PIXE/PIGE

As reported in the experimental section for the PIXE and PIGE, data analysis begins with the summed spectrum for the complete mapped region. In PIXE, maps are generated by fitting the X-ray spectrum, removing the background, and mapping the net counts under the peaks for the elements of interest. In PIGE, elemental maps were generated by determining the net counts under respective peaks after a local linear background subtraction.

A typical X-ray spectrum (PIXE) extracted for the primer is presented in Figure 6a. The position of the X-ray peaks are the same as in normal EDS, since they involve normal K- and L-series lines; however, the lines are generated by proton interaction rather than electron interaction as in normal EDS. The PIXE spectrum of the AA2024-T3 is shown in Figure 6b. In the spectrum from the primer (Figure 6a), the major peaks are Ti, Ba, Fe, Cu, Zn and Zr. Since the primer includes additives such as TiO$_2$, BaSO$_4$ (and SrSO$_4$ as impurity) and Li$_2$CO$_3$, the Ti, Sr and Ba peaks can be attributed to these species. The Zr may arise from a coating applied to the TiO$_2$, since Al and Zr compounds are used to stabilise the TiO$_2$ particles (Table 1). While the Ba and Ti signals overlap in EDS spectra and maps from the SEM, this effect is considerably reduced in PIXE, because the Ba Kα lines dictate the intensity in the Ba Lα lines in the 5.0–7.5 keV region of the spectrum. Thus, there is only a small residual signal of Ba in the Ti map arising from residual fitting errors.

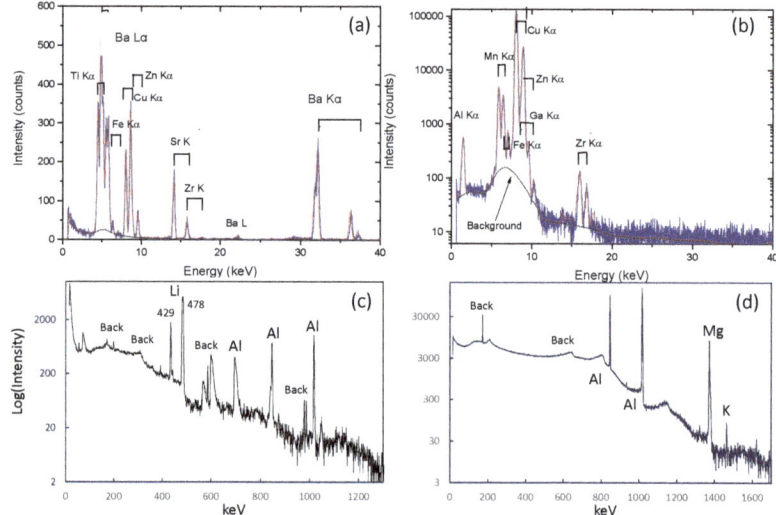

Figure 6. PIXE (X-ray) spectra of the (**a**) primer and (**b**) AA2024-T3. Corresponding PIGE (γ-ray) spectra of the (**c**) primer and (**d**) AA2024-T3. The red lines in (**a**) and (**b**) are the fitted curves to the spectra. The dashed lines in (**a**) and (**b**) are backgrounds used in the fitting.

Individual PIGE spectra for the primer and the AA2024-T3 can be extracted from the maps, and typical examples are shown in Figure 6c,d, respectively. The Y-ray spectrum (PIGE) shows Li (peak positions), Al and Mg. Peaks labelled "back" arise from laboratory background signals and are not part of the sample. The Li peak at 429 keV was used for the determination of the Li distribution. For the AA2024, Figure 6d only shows the Al and Mg signals.

The combined PIXE and PIGE maps for a region of a sample prior to leaching is shown in Figure 7. The Li, Ba, Sr and Ti maps clearly show that these elements are present in the coating. Sr is an impurity in the $BaSO_4$, and is probably present as $SrSO_4$ (Table 1). It should be pointed out that some of these elements are present in very low levels, and it is only through the sensitivity of PIXE that they are detected at all.

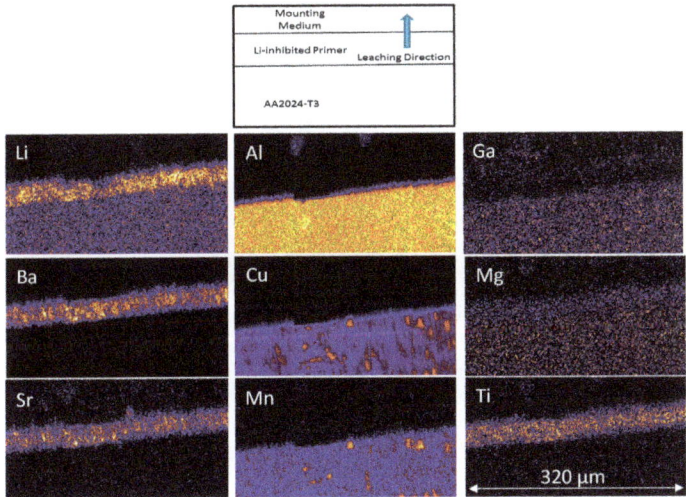

Figure 7. Combined PIXE and PIGE maps for a sample that has not been exposed to NSS. The PIGE maps (Li-PIGE and Al-PIGE) are labelled as such, and the rest are PIXE maps. The schematic on the top shows the sample configuration and indicates the direction and interface from which leaching has occurred. The colour scale is thermal, with warm colours representing higher concentrations.

With respect to the AA2024-T3, the PIXE spectrum of the AA2024-T3 substrate (Figure 6b) shows Al, Cu, Mn, Fe, Zn, Ga and Zr. The Zr may be an additive used in the formation of $ZrAl_3$ used for grain refining [58]. Cu and Mn were detected both in the matrix and constituent IM particles, and Fe only in the constituent particles [59–67]. The presence of Cu and Mn in the matrix can be explained by a small but significant solubility of Cu in Al, as well as Cu and Mn being present in a number of IM particles (hardening precipitates (Cu) and dispersoids ($Al_{20}Mn_3Cu_2$)), which are much smaller than the resolution of the technique [68]. Elements such as Ga have been reported before when using Rutherford Backscattering spectroscopy (RBS) to examine aluminium alloys [69]. In some Al-alloys, Zn is used for precipitate hardening using the η-phase (Zn_2Mg) in 7xxx series alloys [70] but, again, it is not expected as an alloy addition here, even though Zn is detected in the AA2024-T3 sheet product [60]. In this study, it is associated with Cu-containing constituent particles, and may be present as an impurity from a mixed stock starting material used to manufacture the AA2024.

Figure 8 shows three-colour maps of the primer region, where Li is in red and Ba is in blue for all these maps, and green reflects the changing element. The Li-Cu-Ba map indicates the distribution of the Li_2CO_3 (red) and $BaSO_4$ (green) particle distributions within the primer, and the Cu (green) reveals relationship of the primer to the AA2024-T3 substrate. The dark band separating the AA2024-T3 from the primer in the Cu map coincides with a purple strip in the Al map on top of the metal. In the middle and top maps, blue is the anodised layer. In the Li-Sr-Ba map, Ba-containing particles are light blue, indicating a mixing of the colours associated with the Sr (green) with the Ba (blue), which confirms the presence of $SrSO_4$ in the $BaSO_4$. From these maps, it is clear that there are regions that are rich and poor in Li_2CO_3 particles. These regions can be as deep as the coating itself (e.g., point A in Figure 8a) and 20–30 µm wide. There was no suggestion of layering in these maps.

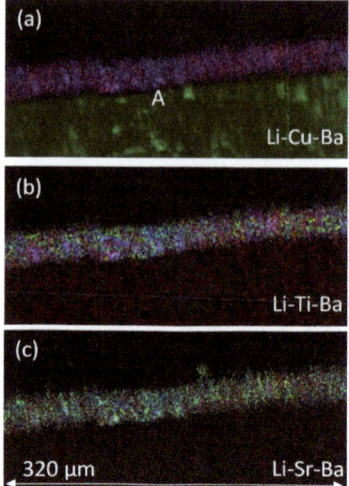

Figure 8. Three-colour maps. (**a**) Li-Cu-Ba, (**b**) Li-Ti-Ba and (**c**) Li-Sr-Ba. In all cases, the Li is red, the barium is blue, the middle element is green; i.e., green changes from top to bottom as Cu, Ti and Sr.

3.2. NSS Exposure and Li Depletion in the Primer

3.2.1. SEM/EDS

As can be seen from Figure 9, exposure to NSS for 500 h resulted in the generation of voids in the primer, which is assumed to be due to the loss of Li_2CO_3 particles. At low magnification, it is difficult to determine the depth of depletion due to the contrast similarity between the Li_2CO_3 particles and voids created by their dissolution. On closer examination of individual sites, however, it was clear that the depth of Li_2CO_3 depletion varied considerably from place to place along the section of the primer. In some places, the removal of Li_2CO_3 particles was from quite deep into the primer to near the metal/primer interface. Examples in Figure 9a shows the presence of voids due to partial/complete dissolution of particles, and is magnified in Figure 9c. These voids appear approximately halfway through the depth of the coating. Figure 9d shows the development of interfacial voids between the Li_2CO_3 particles and the polyurethane matrix, in this instance near the primer/anodised/metal interface. Moreover, channels were often observed at the base of some of these voids, suggesting that they are not isolated, but connected to other voids. The delamination between the inhibitor particles and the polyurethane indicates the possible development of further transport networks within the coating system, as well as changes in chemistry at the interface between these particles and the polyurethane.

Figure 10a shows a backscatter image of a section of the primer with the various inorganic additives as discussed above. The quantitative maps in Figure 10b and c show that there is an absence of large particles near the surface, which only have an oxygen (and carbon) signal and inferred to be Li_2CO_3. The Mg-(hydr)oxide, TiO_2 and $BaSO_4$ distributions appear to be similar to those of the primer without NSS. However, the quantitative analyses show that S and O decrease after NSS exposure of the $BaSO_4$, perhaps suggesting a loss of sulphate ions (the loss of S is roughly 25% the reduction of O) (Table 2). The Mg-(hydr)oxide data shows that the Mg:O ratio has decreased from 1:1.6 to 1:1, suggesting that a mixture of MgO and $Mg(OH)_2$, present prior to NSS, may have been transformed to MgO after NSS exposure. The origin of this transformation is unclear, since MgO is more soluble than $Mg(OH)_2$ under a range of conditions [70], and it would be expected that exposure to the electrolyte would result in an increase in the hydroxide. Lastly, it is worth noting that Cl was detected in the coating after NSS exposure, whereas it was not detected prior to NSS (Figure 11). In the sample after NSS exposure, the chloride appeared to be confined to the polyurethane and was not in either the voids left by the dissolution of the Li$_2$CO$_3$ particles or delamination around them (Figure 11). As, an example the spectrum from the region indicated by the circle in Figure 10a is shown in Figure 11. The implications of these results will be discussed in more detail below.

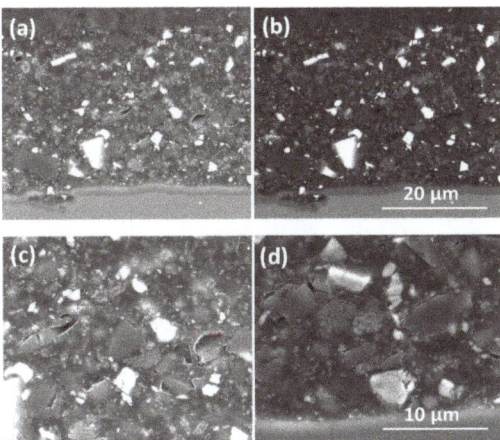

Figure 9. (a) Secondary and (b) backscattered electron images of voids resulting from the dissolution of Li$_2$CO$_3$ inhibitor particles after 500 h NSS exposure; (c) Magnification of (b); (d) Interfacial voids between the primer and the Li$_2$CO$_3$ inhibitor particles deep in the primer near the primer/anodised layer interface. Dashed box in (a) indicates the region in (c).

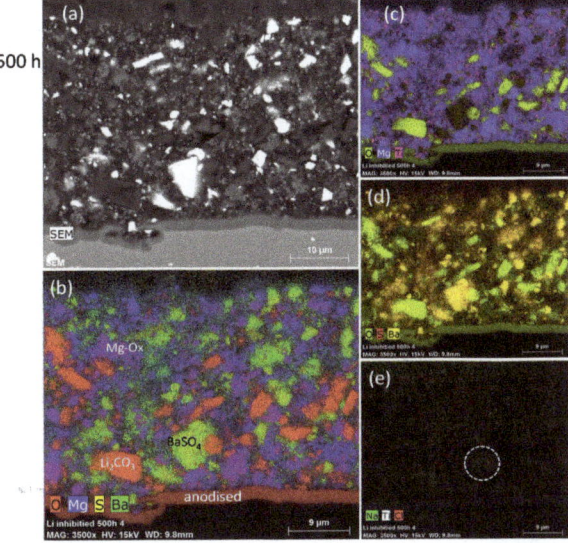

Figure 10. EDS mapping results for sample with 500 h exposure to NSS. (a) Backscattered electron image and maps derived from quantitative analysis and composite maps for (b) O-Mg-Ba-S with phase labelling (c) O-Mg-Ti, (d) O-S-Ba and (e) Na-Ti-Cl map. (N.B. colour mixing is not the same as three-colour mixing).

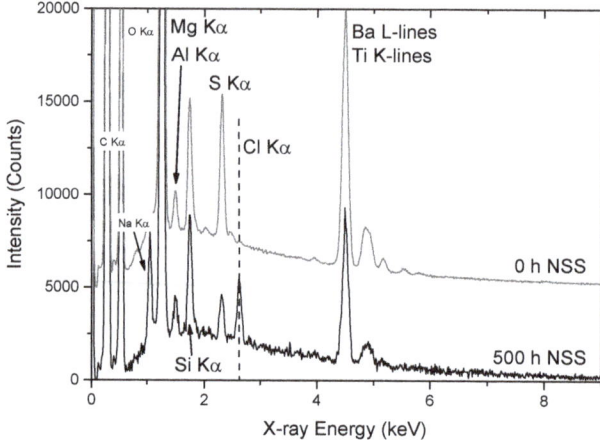

Figure 11. Sample spectra from X-sections of the polyurethane coatings without exposure to (from Figure 5) and 500 h exposure to NSS (Figure 10), respectively. The points from which the spectra have been taken are indicated in each figure by the dashed circle.

3.2.2. PIXE/PIGE

PIXE and PIGE maps are presented in Figure 12 for Li, Sr and Ba sections of the samples prior to NSS and after 500 h NSS. The PIGE results represent an average over approximately 60 µm depth perpendicular to the X-section of the coating, and thus the average over several particle diameters. The maps show that Li is locally concentrated into small regions typically 2–10 µm in size, reflecting the particle size distribution. There is also some suggestion that these features are themselves interconnected, since the smaller features tend to form larger extended structures, suggesting the presence of clusters of Li_2CO_3 particles [71]. In the sample that has not been exposed to NSS, there is a clean boundary between the primer and the mounting medium. After 500 h NSS exposure, there is a zone completely depleted of Li_2CO_3 particles in the surface of the primer, as determined by comparing the Li distribution with the Ba distribution, and indicated by the parallel white lines. The depth of this depleted zone is 11 ± 8 µm (Table 4). However, there are also regions where there is a local, selective removal of individual particles or clusters of particles of Li_2CO_3 that penetrate much deeper into the coating. While this cannot be directly confirmed from the PIXE/PIGE, since the Li_2CO_3 distribution prior to NSS exposure is unknown for any particular region, the SEM clearly shows voids quite deep within the coating, suggesting selective dissolution paths (Figure 9).

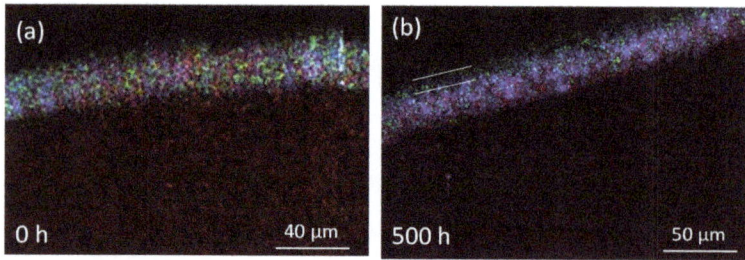

Figure 12. Three-colour maps of sections of the coating prior to NSS exposure and after 500 h exposure to NSS. These maps are combined PIGE (Li–Red) and PIXE, (Sr in Blue and Ba in green). Three colour elemental maps for Li-Cu-Ba (**a**) without NSS exposure and (**b**) after 500 h NSS exposure.

3.3. Measurement of Li-Depletion

The above data show that Li depletion manifests itself in several separate ways, meaning that there isn't a single metric to describe it. Thus, the depletion depth needs to be determined by inference from the different types of data collected here. In this section, three different approaches to measuring depletion depth are presented.

The first method is based on elemental line profiles across the coating using traverse profiles by analysis of the PIXE/PIGE data as displayed in Figure 13. These are constructed by first choosing a region for the profile, as shown by the green box in Figure 13a. The average counts along a number of lines (e.g., the red line) at fixed spacing are then determined to produce a point (red point) in the line profile. Thus integrated elemental profiles for Ti, Ba, Li and Al across sections of the primer are generated and presented in Figure 13b for the sample without exposure to NSS and that exposed for 500 h (Figure 13c). In this case, Ba and Ti profiles were used as indicators of the coating thickness based on the assumption that their distributions were unaffected by NSS exposure. (The thicknesses determined using the Ti maps tended to be slightly larger than those determined from the Ba maps, which may be due to the Ti particles being finer and closer to the surface of the film). The width of the profiles shown in Figure 13 are larger than would be obtained from individual line profiles, because the traverse method averages along a line (red line) at a particular depth through the cross section of the primer, and thus incorporates surface variation, such as roughness. Figure 13b shows Ti, Li, Ba and Al traverse profiles for the sample without NSS exposure. The average thickness was 33.7 ± 2.4 µm, which was determined from five measurements across the coating. (Note: This is different from the SEM measurement since it is a different region of the sample.) In repeated measurements, the Li profiles showed that there was a region near the surface where there was a lower concentration of Li for the sample without NSS exposure. Such a region is indicated in Figure 13b and labelled the "skin layer". This may be due to fewer smaller Li_2CO_3 particles compared to the other additives (Table 2). The thickness of this zone of lower concentration (skin layer) appeared to be in the vicinity of 3–5 µm, as determined from line scans (not the traverse method).

The thickness for the sample exposed for 500 h to NSS was 45.4 ± 6.5 µm, which was determined in the same fashion as the sample without NSS exposure. In this instance, the depleted zone near the surface was in the vicinity of 11 µm (Table 4). As stated in the experimental section, the actual Li concentrations are difficult to determine in a complex matrix such as the polyurethane with a heterogeneous distribution of inorganic additives. However, in Figure 13c the Al PIGE profiles have been adjusted to have the same level of counts in the metal, which allows a *qualitative* comparison of the Li profiles. The position of the surface is indicated for both profiles, remembering that the two samples have different coating thicknesses. It can be seen that there is depletion of Li from a greater depth into the primer for the coating exposed to NSS for 500 h. Moreover, there is significant Li depletion from the body of the coating to around 30% of the level in the sample without NSS, indicating a depletion of Li from within the coating. This is probably due to dissolution of the surface on Li_2CO_3 particles, which are deeper in the coating.

Table 4. Characteristic thickness of Li-depletion zones in microns.

Sample	Total Thickness (µm)	Skin Layer (µm)	Homogenously Depleted Zone (µm)	Deepest Depletion (µm)
SEM No NSS	33.5 ± 5	3–4	–	–
PIXE/PIGE No NSS	33.7 ± 2.5	Up to 5	–	–
PIXE/PIGE 500 h NSS	45.4 ± 6.5	–	11 ± 8	25
SEM 500 h NSS	32.6 ± 1.0	–	9–12	Film thickness

A second method for determining the depletion depth was applied based on the local absence of Li from the PIGE Li map of the cross section of the film. In this method, the local depletion of Li was assessed compared to the thickness of the primer coating, as shown in Figure 14a (white dashed line).

A drawback of this method is that the Li_2CO_3 distribution prior to the NSS exposure is unknown, so, while the absence Li_2CO_3 particles in any particular region of the primer coating is assumed to be due to inhibitor loss, in reality it will include regions where there were no Li_2CO_3 particles to start with. As expected, the depletion measured using this method shows a larger zone of depletion for the 500 h NSS exposure sample than the sample without exposure to NSS ("SEM 500 h NSS" under "Deepest Depletion" Table 4). The absence of Li at the top of the coating with 500 h NSS exposure is inferred from the EDS measurements presented in Figure 10b. This provides the second method for depletion thickness determination being in the region 9–12 µm, which is similar to that measured using the PIGE traverse approach.

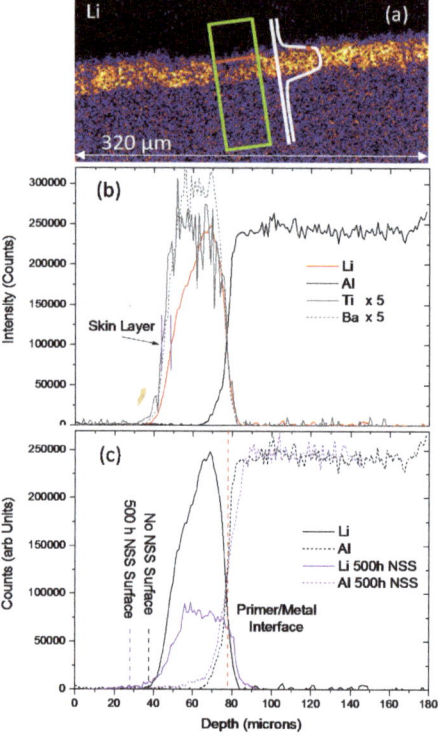

Figure 13. Traverse elemental profiles. (**a**) Example of a traverse profile where a region is chosen for the profile (green box) and individual elemental intensities are averaged along a line (red line) to produce a point on a line profile (red point); (**b**) Profiles for Ti (PIXE), Ba (PIXE), Li (PIGE) and Al (PIGE) across the without NSS. Ti and Ba signals are multiplied by five to show the skin layer. (**c**) PIGE profiles for Al and Li without salt spray and after 500 h NSS. Note the thickness of the coatings are different and the surface is indicated by markers. The 500 h NSS results have been adjusted so that the Al intensity from both conditions is the same, thus allowing *qualitative* comparison between the two Li profiles.

A third approach is to use SEM to determine the deepest point where there are voids (assumed to be due to Li_2CO_3 dissolution) in the coating. An example of void distribution for the sample exposed for 500 h to NSS is shown in Figure 14b, and voids can also be seen in Figure 10a. Summary depletion depths for the deepest depletion are presented in Table 4. Figure 14c is a higher magnification image of the region near the surface showing the extent of interconnection between the voids, which indicate a cluster formation, as previously reported by Hughes et al. for chromate clusters in an epoxy-based coating [72–74]. These data show that local depletion within the coating can be considerably deeper than the homogeneous depletion depth.

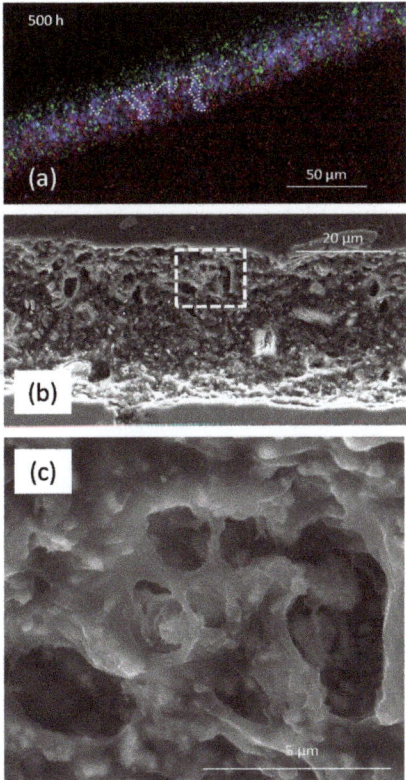

Figure 14. (a) Example of the local dissolution front on the three-colour map of the sample exposed to NSS for 500 h; (b) Secondary electron image of the distribution of voids within a region of the sample exposed to NSS for 500 h; (c) Enlargement of the region within the square in (b) showing the connectivity between the voids. The voids are assumed to be due to the dissolution and loss of Li_2CO_3 particles.

4. Discussion

In this paper, the combination of SEM/EDS and PIXE/PIGE have been used to investigate the distribution and chemical composition of inorganic components in a polyurethane coating prior to and after NSS exposure. The SEM/EDS results show that particle sizes observed in the coating are similar to those measured on free particles. They also show that the smaller particles of each type tend to be more homogenously dispersed than the larger particles. In the case of Li_2CO_3, which has fewer small particles, this appears to result in a "skin layer" near the surface where there are far more smaller TiO_2, $BaSO_4$ and Mg(hydr)oxide particles.

It was shown that (semi-) quantitative analysis can be used to separate the distribution of TiO_2 from $BaSO_4$ in SEM/EDS, and in PIXE these are separated using the Ba Kα line. PIXE was also able to identify the presence of Sr in $BaSO_4$. Upon exposure to NSS, SEM/EDS was able to provide useful information on the loss of Li_2CO_3, even though Li could not be directly detected. This was achieved using a combination of PIXE and PIGE to show that there was loss of Li accompanied by a change in the morphology of the coating through the formation of voids via dissolution of Li_2CO_3 particles. Moreover, there were other changes in the coating, such as the change in the Mg:O ratio of the Mg-(hydr)oxide upon NSS exposure.

The results presented here show that the determination of a Li_2CO_3 depletion depth in organic coatings is complicated by the types of technique available to detect Li, as well as the morphology of the depletion itself. In the former case, none of the techniques presented here can be individually be used to determine the level of leaching. In the case of the PIGE, while it is possible to map the Li distribution and show that there is a region of homogeneous depletion as well as possible local depletion, this data needs to be supplemented by SEM and PIXE. In the case of SEM, it is necessary to ensure for regions of apparent depletion that voids, which indicate the depletion, have been created through the dissolution of Li_2CO_3 particles. This is to distinguish leaching phenomena from inhomogeneous distributions of Li_2CO_3 particles. PIXE is also required to supplement the PIGE results, since other inorganics in the coating act as markers for the coating thickness. Examples include the Ba and Ti distributions, which have been used here to determine the coating thickness.

The combination of PIGE/PIXE and SEM/EDS reveal that depletion of Li from the Li_2CO_3 loaded primer after 500 h NSS exposure is a complex process. First, there is the development of a uniform depletion zone from the surface. The changes to Mg-(hydr)oxide may lead to additional pathways for the release of Li_2CO_3 that has not been released up to that point. The presence of local depletion of Li_2CO_3 penetrating further into the coating below this zone, rather than uniform depletion, suggests that clusters of Li_2CO_3 particles are involved in the release process. Release from a cluster involves (i) direct connection of the cluster to the external electrolyte and (ii) gradual dissolution from the cluster. The gradual dissolution need not necessarily move as a "front" through the cluster (i.e., particles nearer the surface must completely dissolve before the next particles can dissolve), but may occur through the simultaneous dissolution of particles at different depths in the clusters, which is dictated by transport of the inhibitor through the electrolyte in the cluster/void structure. The creation of these voids generates a fractal network that acts a porous medium. Small voids and delamination from the polyurethane observed around Li_2CO_3 particles support this hypothesis. The detailed dynamics of the release would need to be determined as a function of time and cannot be revealed here, since only two times were examined. These concepts are summarised in Figure 15. It should be noted that in the chromate case, these pathways were important because the size of the chromate ion meant that it could not diffuse through the epoxy, only through channels created by the dissolution of the chromate particles. In the case of Li_2CO_3 dissolution, while it is possible that the Li ion might be small enough to diffuse through the polyurethane by itself, it is much more likely to diffuse through channels connected to the external electrolyte and created by the dissolution of clusters of Li_2CO_3 particles themselves.

Finally, Figures 5e and 10e show a Ti-Cl and a Ti-Cl-Na map for 0 h NSS and 500 h NSS, respectively. Neither Na nor Cl was detected in the primer coating without exposure to NSS; however, both were detected in the sample exposed to NSS for 500 h (Figure 10). The intensity of Na and Cl varied across the sample exposed to 500 h NSS, which suggested separate pathways for the diffusion of each of these ions. It was not significant in any of the larger inorganic particles, and only appeared in the polyurethane matrix. Significantly, it was not observed near channels created by the dissolution of the Li_2CO_3. In a model where the leaching is via transport paths created by the dissolution of the Li_2CO_3 particles, the role of the external electrolyte needs to be considered. The voids created by the interfacial interaction of inhibitor particles or their complete dissolution appear to be clean in the SEM studies presented here. This suggests that there must be liquid in the voids that is lost during the sample preparation process, for if there were precipitates in these voids, the preparation procedure would capture it (non-polar solvents were used for preparation, so dissolution of precipitates is unlikely). Na and Cl were only detected away from the Li_2CO_3 particles, either in the polyurethane matrix or in other pathways originating from interfaces between the polyurethane and the non-inhibitor inorganic components. This suggests that only water without the salt components (Na or Cl) diffuses into the inhibitor pathways. This is probably because these pathways, generated by Li_2CO_3 dissolution, quickly become saturated with ions from the inhibitor phase, providing an ionic barrier to ions of the external electrolyte. It could also be concluded from these observations that the transport of electrolytes through the coating is complex, with multiple and separate pathways for the external electrolytes and the "internal electrolytes" (water with inhibitor ions): an area that warrants further investigation.

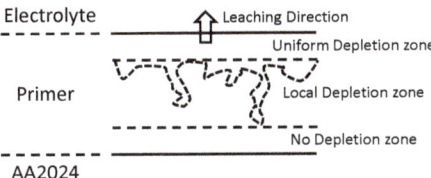

Figure 15. Model of leaching from the Li_2O_3-inhibited primer based on the observations for 500 h exposure to NSS. Leaching appears to occur both uniformly as well as locally. The uniform depletion appears to be associated with changes in coating, whereas the local depletion appears to be associated with selective removal of Li_2O_3 particles.

5. Conclusions

PIGE, PIXE and SEM/EDS have been used to study the distribution and chemistry of inorganic components in a polyurethane coating prior to and after 500 h NSS exposure. Prior to NSS exposure, the PIGE results revealed that there was a zone (3–4 µm deep) near the surface of the primer that appeared to have less Li_2CO_3, thus forming a type of "skin" layer. This "skin" layer had smaller particles of TiO_2 and $BaSO_4$. After NSS exposure, the PIGE results indicated that there was a homogeneously depleted zone extending from the electrolyte/primer interface into the primer (in this case around 11 µm), plus local depletion penetrating much deeper into the primer with a maximum measured depth of around 25 µm. The deeper local penetration was confirmed using SEM, where voids created by partial and complete dissolution of Li_2CO_3 particles were observed extending deep into the primer towards the metal/primer interface. Magnesium hydroxide/oxide particles also appeared to undergo some change with exposure to NSS, with the Mg:O ratio moving closer to 1. The other inorganic particles (TiO_2 and $BaSO_4$) appeared unchanged upon NSS exposure. There was some evidence of chloride penetration into the polyurethane component of the primer, but not within the channels created around the Li_2CO_3 particles.

Author Contributions: Peter Visser, Arjan Mol and Herman Terryn conceived and designed the experiments. Peter Visser prepared samples and performed the NSS testing. Herman Terryn, Arjan Mol, Peter Visser and Tony Hughes were involved in the interpretation of data. Jamie Laird performed the nuclear microprobe experiments and was assisted by Tony Hughes and Chris Ryan in data analysis and interpretation. Tony Hughes performed all SEM experiments and interpreted the data. Tony Hughes wrote the paper with input from all other authors.

Conflicts of Interest: There are no conflicts of interest.

Appendix A

The influence of subsurface particles can be seen in Figure A1, which shows images of the same region collected at different accelerating voltages. The ellipses, highlighted using dashed lines in Figure A1a, show two regions where, with increasing voltage, subsurface $BaSO_4$ particles become evident. Thus, the contribution from subsurface particles may distort the analysis of small particles.

EDS maps for a section of the primer are displayed in Figure A2. These maps highlight the difficulty of using EDS mapping alone to study the distributions (and redistribution after leaching) of all the phases that comprise the primer coating. Elements such as Cu and Fe result from changes in the background level in the spectral region of the Kα lines of these transition metals. Of course, the presence of an element in a particular region of the map can and should be checked using EDS spectra where a change in background is easily distinguished from a peak. However, overlapping lines are a different issue. This is important from the perspective of following the changes in primer additives, such as with Ti and Ba, where overlapping lines give misleading information on the distribution of these particles. In this case, the overlapping lines show a co-incidence of Ti- and Ba-containing particles on their respective maps (Figure A2). The overlap in the X-ray lines themselves is clearly seen in Figure 4, where the Ti Kα lines overlap with the Ba L-series lines, and the Si Kα overlaps with the Sr Kα lines. This can only be resolved using a fitting of the EDS spectra, such as that which is achieved when quantifying the spectra. This approach has been used in this paper.

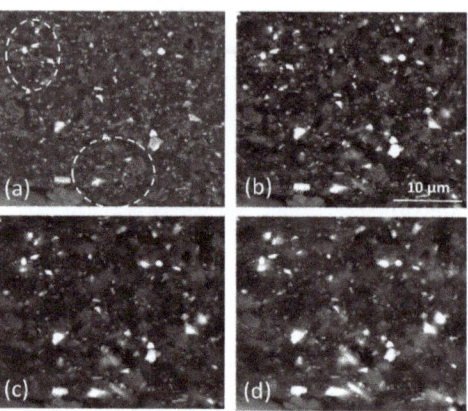

Figure A1. Backscatter electron images collected at (**a**) 10 kV, (**b**) 15 kV, (**c**) 20 kV and (**d**) 25 kV. Dashed ellipses highlight regions where the backscatter contrast changes significantly with accelerating voltage.

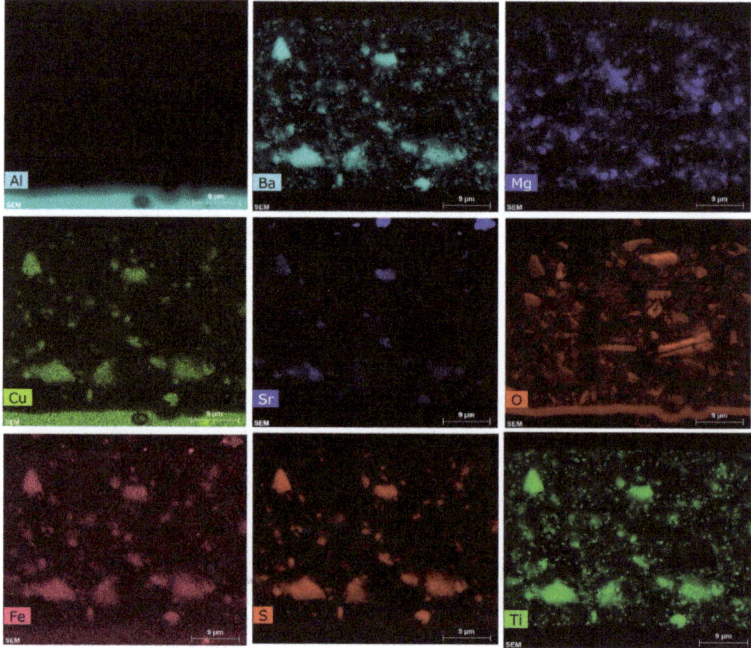

Figure A2. Elemental maps collected from the section of the primer prior to exposure to NSS.

References

1. Catubig, R.; Hughes, A.E.; Cole, I.S.; Hinton, B.R.W.; Forsyth, M. The use of cerium and praseodymium mercaptoacetate as thiol-containing inhibitors for AA2024-T3. *Corros. Sci.* **2014**, *81*, 45–53. [CrossRef]
2. Markley, T.A.; Mardel, J.I.; Hughes, A.E.; Hinton, B.R.W.; Glenn, A.M.; Forsyth, M. Chromate replacement in coatings for corrosion protection of aerospace aluminium alloys. *Mater. Corros. Werkstoffe Und Korros.* **2011**, *62*, 836–840. [CrossRef]
3. Hughes, A.E.; Ho, D.; Forsyth, M.; Hinton, B.R.W. Towards replacement of chromate inhibitors by rare earth systems. *Corros. Rev.* **2007**, *25*, 591–605. [CrossRef]
4. Mardel, J.; Garcia, S.J.; Corrigan, P.A.; Markley, T.; Hughes, A.E.; Muster, T.H.; Lau, D.; Harvey, T.G.; Glenn, A.M.; White, P.A.; et al. The characterisation and performance of Ce(dbp)3-inhibited epoxy coatings. *Prog. Org. Coat.* **2011**, *70*, 91–101. [CrossRef]
5. Hughes, A.E.; Cole, I.S.; Muster, T.H.; Varley, R.J. Combining Green and Self Healing for a new Generation of Coatings for Metal Protection. *Nat. Asia Mater.* **2010**, *2*, 143–151. [CrossRef]
6. Shi, H.; Han, E.-H.; Lamaka, S.V.; Zheludkevich, M.L.; Liu, F.; Ferreira, M.G.S. Cerium cinnamate as an environmentally benign inhibitor pigment for epoxy coatings on AA 2024-T3. *Prog. Org. Coat.* **2014**, *77*, 765–773. [CrossRef]
7. Yasakau, K.A.; Tedim, J.; Zheludkevich, M.L.; Drumm, R.; Shem, M.; Wittmar, M.; Veith, M.; Ferreira, M.G.S. Cerium molybdate nanowires for active corrosion protection of aluminium alloys. *Corros. Sci.* **2012**, *58*, 41–51. [CrossRef]
8. Schem, M.; Schmidt, T.; Gerwann, J.; Wittmar, M.; Veith, M.; Thompson, G.E.; Molchan, I.S.; Hashimoto, T.; Skeldon, P.; Phani, A.R.; et al. CeO_2-filled sol-gel coatings for corrosion protection of AA2024-T3 aluminium alloy. *Corros. Sci.* **2009**, *51*, 2304–2315. [CrossRef]
9. Yasakau, K.A.; Zheludkevich, M.L.; Ferreira, M.G.S. Lanthanide salts as corrosion inhibitors for AA5083. Mechanism and efficiency of corrosion inhibition. *J. Electrochem. Soc.* **2008**, *155*, C169–C177. [CrossRef]
10. Montemor, M.F.; Trabelsi, W.; Lamaka, S.V.; Yasakau, K.A.; Zheludkevich, M.L.; Bastos, A.C.; Ferreira, M.G.S. The synergistic combination of bis-silane and $CeO_2.ZrO_2$ nanoparticles on the electrochemical behaviour of galvanised steel in NaCl solutions. *Electrochim. Acta* **2008**, *53*, 5913–5922. [CrossRef]
11. Zheludkevich, M.L.; Serra, R.; Montemor, M.F.; Yasakau, K.A.; Salvado, I.M.M.; Ferreira, M.G.S. Nanostructured sol-gel coatings doped with cerium nitrate as pre-treatments for AA2024-T3—Corrosion protection performance. *Electrochim. Acta* **2005**, *51*, 208–217. [CrossRef]
12. Paussa, L.; Andreatta, F.; de Felicis, D.; Bemporad, E.; Fedrizzi, L. Investigation of AA2024-T3 surfaces modified by cerium compounds: A localized approach. *Corros. Sci.* **2014**, *78*, 215–222. [CrossRef]
13. Andreatta, F.; Druart, M.E.; Lanzutti, A.; Lekka, M.; Cossement, D.; Olivier, M.G.; Fedrizzi, L. Localized corrosion inhibition by cerium species on clad AA2024 aluminium alloy investigated by means of electrochemical micro-cell. *Corros. Sci.* **2012**, *65*, 376–386. [CrossRef]
14. Paussa, L.; Rosero-Navarro, N.C.; Andreatta, F.; Castro, Y.; Duran, A.; Aparicio, M.; Fedrizzi, L. Inhibition effect of cerium in hybrid sol-gel films on aluminium alloy AA2024. *Surf. Interface Anal.* **2010**, *42*, 299–305. [CrossRef]
15. Ralston, K.D.; Young, T.L.; Buchheit, R.G. Electrochemical Evaluation of Constituent Intermetallics in Aluminum Alloy 2024-T3 Exposed to Aqueous Vanadate Inhibitors. *J. Electrochem. Soc.* **2009**, *156*, C135–C146. [CrossRef]
16. Ralston, K.D.; Chrisanti, S.; Young, T.L.; Buchheit, R.G. Corrosion inhibition of aluminum alloy 2024-T3 by aqueous vanadium species. *J. Electrochem. Soc.* **2008**, *155*, C350–C359. [CrossRef]
17. Harvey, T.G.; Hardin, S.G.; Hughes, A.E.; Muster, T.H.; White, P.A.; Markley, T.A.; Corrigan, P.A.; Mardel, J.; Garcia, S.J.; Mol, J.M.C.; et al. The effect of inhibitor structure on the corrosion of AA2024 and AA7075. *Corros. Sci.* **2011**, *53*, 2184–2190. [CrossRef]
18. Zheludkevich, M.L.; Tedim, J.; Freire, C.S.R.; Fernandes, S.C.M.; Kallip, S.; Lisenkov, A.; Gandini, A.; Ferreira, M.G.S. Self-healing protective coatings with "green" chitosan based pre-layer reservoir of corrosion inhibitor. *J. Mater. Chem.* **2011**, *21*, 4805–4812. [CrossRef]
19. Raps, D.; Hack, T.; Wehr, J.; Zheludkevich, M.L.; Bastos, A.C.; Ferreira, M.G.S.; Nuyken, O. Electrochemical study of inhibitor-containing organic-inorganic hybrid coatings on AA2024. *Corros. Sci.* **2009**, *51*, 1012–1021. [CrossRef]

20. Montemor, M.F. Fostering green inhibitors for corrosion prevention. In *Series in Materials Science*; Springer: Berlin, Germany, 2016; pp. 107–137.
21. El-Faham, A.; Dahlous, K.A.; al Othman, Z.A.; Al-Lohedan, H.A.; El-Mahdy, G.A. sym-Trisubstituted 1,3,5-Triazine Derivatives as Promising Organic Corrosion Inhibitors for Steel in Acidic Solution. *Molecules* **2016**, *21*, 436. [CrossRef] [PubMed]
22. Winkler, D.A.; Breedon, M.; Hughes, A.E.; Burden, F.R.; Barnard, A.S.; Harvey, T.G.; Cole, I. Towards chromate-free corrosion inhibitors: Structure-property models for organic alternatives. *Green Chem.* **2014**, *16*, 3349–3357. [CrossRef]
23. Gonzalez-Olvera, R.; Roman-Rodriguez, V.; Negron-Silva, G.E.; Espinoza-Vazquez, A.; Rodriguez-Gomez, F.J.; Santillan, R. Multicomponent Synthesis and Evaluation of New 1,2,3-Triazole Derivatives of Dihydropyrimidinones as Acidic Corrosion Inhibitors for Steel. *Molecules* **2016**, *21*, 250. [CrossRef] [PubMed]
24. Allahar, K.N.; Wang, D.; Battocchi, D.; Bierwagen, G.P.; Balbyshev, S. Real-Time Monitoring of a United States Air Force Topcoat/Mg-Rich Primer System in ASTM B117 Exposure by Embedded Electrodes. *Corros. Sci.* **2010**, *66*, 075003. [CrossRef]
25. Bierwagen, G.; Brown, R.; Battocchi, D.; Hayes, S. Active metal-based corrosion protective coating systems for aircraft requiring no-chromate pretreatment. *Prog. Org. Coat.* **2010**, *68*, 48–61. [CrossRef]
26. Wang, D.H.; Battocchi, D.; Allahar, K.N.; Balbyshev, S.; Bierwagen, G.P. In situ monitoring of a Mg-rich primer beneath a topcoat exposed to Prohesion conditions. *Corros. Sci.* **2010**, *52*, 441–448. [CrossRef]
27. Xu, H.; Battocchi, D.; Tallman, D.E.; Bierwagen, G.P. Use of Magnesium Alloys as Pigments in Magnesium-Rich Primers for Protecting Aluminum Alloys. *Corros. Sci.* **2009**, *65*, 318–325. [CrossRef]
28. Figueira, R.; Fontinha, I.; Silva, C.; Pereira, E. Hybrid Sol-Gel Coatings: Smart and Green Materials for Corrosion Mitigation. *Coatings* **2016**, *6*, 12. [CrossRef]
29. Kozmel, T.; Vural, M.; Tin, S. EBSD analysis of high strain rate application Al-Cu based alloys. *Mater. Sci. Eng. A* **2015**, *630*, 99–106. [CrossRef]
30. Zheludkevich, M.L.; Poznyak, S.K.; Rodrigues, L.M.; Raps, D.; Hack, T.; Dick, L.F.; Nunes, T.; Ferreira, M.G.S. Active protection coatings with layered double hydroxide nanocontainers of corrosion inhibitor. *Corros. Sci.* **2010**, *52*, 602–611. [CrossRef]
31. Snihirova, D.; Lamaka, S.V.; Taryba, M.; Salak, A.M.; Kallip, S.; Zheludkevich, M.L.; Ferreira, M.G.S.; Montemor, M.F. Hydroxyapatite microparticles as feedback-active reservoirs of corrosion inhibitors. *ACS Appl. Mater. Interfaces* **2010**, *2*, 3011–3022. [CrossRef] [PubMed]
32. Poznyak, S.K.; Tedim, J.; Rodrigues, L.M.; Salak, A.N.; Zheludkevich, M.L.; Dick, L.F.P.; Ferreira, M.G.S. Novel Inorganic Host Layered Double Hydroxides Intercalated with Guest Organic Inhibitors for Anticorrosion Applications. *ACS Appl. Mater. Interfaces* **2009**, *1*, 2353–2362. [CrossRef] [PubMed]
33. Mahajanarn, S.P.V.; Buchheit, R.G. Characterization of inhibitor release from Zn-Al-[$V_{10}O_{28}$]$^{6-}$ hydrotalcite pigments and corrosion protection from hydrotalcite-pigmented epoxy coatings. *Corros. Sci.* **2008**, *64*, 230–240. [CrossRef]
34. Chrisanti, S.; Buchheit, R.G. Use of Ce-modified bentonite clay as a pigment for corroion inhibition and sensing. In *Papers Presented at the Philadelphia, Pennsylvania Meeting*; American Chemical Society: Washington, DC, USA, 2004; Volume 45, pp. 148–149.
35. Mahajanam, S.P.V.; Buchheit, R.G. Characterization of Zn-Al-$V_{10}O_{28}^{6-}$ corrosion-inhibiting hydrotalcite pigments in epoxy resins. In *Corrosion and Protection of Light Metal Alloys*; Buchheit, R.G., Kelly, R.G., Missert, N.A., Shaw, B.A., Eds.; Electrochemical Society Inc.: Pennington, NJ, USA, 2004; pp. 270–282.
36. Buchheit, R.G.; Guan, H. Formation and characteristics of Al-Zn hydrotalcite coatings on galvanized steel. *JCT Res.* **2004**, *1*, 277–290. [CrossRef]
37. Kendig, M.; Hon, M. A hydrotalcite-like pigment containing an organic anion corrosion inhibitor. *Electrochem. Solid State Lett.* **2005**, *8*, B10–B11. [CrossRef]
38. McMurray, H.N.; Williams, G. Inhibition of filiform corrosion on organic-coated aluminum alloy by hydrotalcite-like anion-exchange pigments. *Corros. Sci.* **2004**, *60*, 219–228. [CrossRef]
39. Williams, G.; McMurray, H.N. Inhibition of filiform corrosion on polymer coated AA2024-T3 by hydrotalcite-like pigments incorporating organic anions. *Electrochem. Solid State Lett.* **2004**, *7*, B13–B15. [CrossRef]
40. Williams, G.; McMurray, H.N. Anion-exchange inhibition of filiform corrosion on organic coated AA2024-T3 aluminum alloy by hydrotalcite-like pigments. *Electrochem. Solid State Lett.* **2003**, *6*, B9–B11. [CrossRef]

41. Williams, G.; McMurray, H.N.; Worsley, D.A. Cerium(III) inhibition of corrosion-driven organic coating delamination studied using a scanning kelvin probe technique. *J. Electrochem. Soc.* **2002**, *149*, B154–B162. [CrossRef]
42. Visser, P.; Terryn, H.; Mol, J.M.C. Aerospace Coatings. *Act. Prot. Coat.* **2016**, *233*, 315–372.
43. Visser, P.; Liu, Y.; Zhou, X.; Hashimoto, T.; Thompson, G.E.; Lyon, S.B.; van der Ven, L.G.J.; Mol, A.J.M.C.; Terryn, H.A. The corrosion protection of AA2024-T3 aluminium alloy by leaching of lithium-containing salts from organic coatings. *Faraday Discuss.* **2015**, *180*, 511–526. [CrossRef] [PubMed]
44. Visser, P.; Lutz, A.; Mol, J.M.C.; Terryn, H. Study of the formation of a protective layer in a defect from lithium-leaching organic coatings. *Prog. Org. Coat.* **2016**, *99*, 80–90. [CrossRef]
45. Visser, P.; Liu, Y.; Terryn, H.; Mol, J.M.C. Lithium salts as leachable corrosion inhibitors and potential replacement for hexavalent chromium in organic coatings for the protection of aluminum alloys. *J. Coat. Technol. Res.* **2016**, *13*, 557–566. [CrossRef]
46. Liu, Y.; Visser, P.; Zhou, X.; Lyon, S.B.; Hashimoto, T.; Curioni, M.; Gholinia, A.; Thompson, G.E.; Smyth, G.; Gibbon, S.R.; et al. Protective film formation on AA2024-T3 Aluminum Alloy by leaching of lithium carbonate from an organic coating. *J. Electrochem. Soc.* **2016**, *163*, C45–C53. [CrossRef]
47. Laird, J.S.; Hughes, A.E.; Ryan, C.G.; Visser, P.; Terryn, H.; Mol, J.M.C. Particle induced gamma and X-ray emission spectroscopies of lithium based alloy coatings. *Nucl. Instrum. Methods Phys. Res. Sect. B Beam Interact. Mater. Atoms* **2017**, *404*, 167–172. [CrossRef]
48. Boag, A.; Taylor, R.J.; Muster, T.H.; Goodman, N.; McCulloch, D.; Ryan, C.; Rout, B.; Jamieson, D.; Hughes, A.E. Stable pit formation on AA2024-T3 in a NaCl environment. *Corros. Sci.* **2010**, *52*, 90–103. [CrossRef]
49. Boag, A.P.; McCulloch, D.G.; Jamieson, D.N.; Hearne, S.M.; Hughes, A.E.; Ryan, C.G.; Toh, S.K. Combined nuclear microprobe and TEM study of corrosion pit nucleation by intermetallics in aerospace aluminium alloys. *Nucl. Instrum. Methods Phys. Res. Sect. B Beam Interact. Mater. Atoms* **2005**, *231*, 457–462. [CrossRef]
50. Furman, S.A.; Scholes, F.H.; Hughes, A.E.; Jamieson, D.N.; Macrae, C.M.; Glenn, A.M. Corrosion in artificial defects. II. Chromate reactions. *Corros. Sci.* **2006**, *48*, 1827–1847. [CrossRef]
51. Polmear, I.J. *Light Alloys: Metallurgy of the Light Metals*, 3rd ed.; Arnold: London, UK, 1995.
52. *ASTM B117-16 Standard Practice for Operating Salt Spray (Fog) Apparatus*; ASTM International: West Conshohocken, PA, USA, 2016.
53. Ryan, C.G.; Jamieson, D.N.; Griffin, W.L.; Cripps, G.; Szymanski, R. The new CSIRO-GEMOC nuclear microprobe: First results, performance and recent applications. *Nucl. Instrum. Methods Phys. Res. Sect. B Beam Interact. Mater. Atoms* **2001**, *181*, 12–19. [CrossRef]
54. Laird, J.S.; Szymanski, R.; Ryan, C.G.; Gonzalez-Alvarez, I. A Labview based FPGA data acquisition with integrated stage and beam transport control. *Nucl. Instrum. Methods Phys. Res. Sect. B Beam Interact. Mater. Atoms* **2013**, *306*, 71–75. [CrossRef]
55. Ryan, C.G.; van Achterbergh, E.; Yeats, C.J.; Tin Win, T.; Cripps, G. Quantitative PIXE trace element imaging of minerals using the new CSIRO-GEMOC Nuclear Microprobe. *Nuclear Instrum. Methods Phys. Res. Section B Beam Interact. Mater. Atoms* **2002**, *189*, 400–407. [CrossRef]
56. Boni, C.; Cereda, E.; Marcazzan, G.M.B.; de Tomasi, V. Prompt gamma emission excitation functions for PIGE analysis. *Nucl. Inst. Methods Phys. Res. B* **1988**, *35*, 80–86. [CrossRef]
57. Abrahami, S.T.; Hauffman, T.; de Kok, J.M.M.; Mol, J.M.C.; Terryn, H. XPS Analysis of the Surface Chemistry and Interfacial Bonding of Barrier-Type Cr(VI)-Free Anodic Oxides. *J. Phys. Chem. C* **2015**, *119*, 19967–19975. [CrossRef]
58. Hatch, J.E. *Aluminium: Porperties and Physical Metallurgy*; ASM International: Materials Park, OH, USA, 1984.
59. Ovcharenko, R.E.; Tupitsyn, I.I.; Savinov, E.P.; Voloshina, E.N.; Dedkov, Y.S.; Shulakov, A.S. Calculation of the X-ray emission K and L 2,3 bands of metallic magnesium and aluminum with allowance for multielectron effects. *J. Exp. Theor. Phys.* **2014**, *118*, 11–17. [CrossRef]
60. Hughes, A.E.; Glenn, A.M.; Wilson, N.; Moffatt, A.; Morton, A.J.; Buchheit, R.G. A consistent description of intermetallic particle composition: An analysis of ten batches of AA2024-T3. *Surf. Interface Anal.* **2013**, *45*, 1558–1563. [CrossRef]
61. Hughes, A.E.; Birbilis, N.; Mol, J.M.C.; Garcia, S.J.; Zhou, X.; Thompson, G.E. High Strength Al-Alloys: Microstructure, Corrosion and Principles of Protection. In *Recent Trends in Processing and Degradation of Aluminium Alloys*; Ahmad, Z., Ed.; Intech Publishing: Rijeka, Croatia, 2011.

62. Hughes, A.E.; MacRae, C.; Wilson, N.; Torpy, A.; Muster, T.H.; Glenn, A.M. Sheet AA2024-T3: A new investigation of microstructure and composition. *Surf. Interface Anal.* **2010**, *42*, 334–338. [CrossRef]
63. Buchheit, R.G.; Grant, R.P.; Hlava, P.F.; McKenzie, B.; Zender, G.L. Local dissolution phenomena associated with S phase (Al_2CuMg) particles in aluminum alloy 2024-T3. *J. Electrochem. Soc.* **1997**, *144*, 2621–2628. [CrossRef]
64. Lacroix, L.; Ressier, L.; Blanc, C.; Mankowski, G. Combination of AFM, SKPFM, and SIMS to study the corrosion behavior of S-phase particles in AA2024-T351. *J. Electrochem. Soc.* **2008**, *155*, C131–C137. [CrossRef]
65. Lacroix, L.; Ressier, L.; Blanc, C.; Mankowski, G. Statistical study of the corrosion behavior of Al2CuMg intermetallics in AA2024-T351 by SKPFM. *J. Electrochem. Soc.* **2008**, *155*, C8–C15. [CrossRef]
66. Ilevbare, G.O.; Schneider, O.; Kelly, R.G.; Scully, J.R. In situ confocal laser scanning microscopy of AA 2024-T3 corrosion metrology—I. Localized corrosion of particles. *J. Electrochem. Soc.* **2004**, *151*, B453–B464. [CrossRef]
67. Schneider, O.; Ilevbare, G.O.; Scully, J.R.; Kelly, R.G. In situ confocal laser scanning microscopy of AA 2024-T3 corrosion metrology—II. Trench formation around particles. *J. Electrochem. Soc.* **2004**, *151*, B465–B472. [CrossRef]
68. Hughes, A.E.; Parvizi, R.; Forsyth, M. Microstructure and corrosion of AA2024. *Corros. Rev.* **2015**, *33*, 1–30. [CrossRef]
69. Wu, X.; Hebert, K. Development of Surface Impurity Segregation during Dissolution of Aluminum. *J. Electrochem. Soc.* **1996**, *143*, 83–91. [CrossRef]
70. Cavanaugh, M.K.; Birbilis, N.; Buchheit, R.G. Modeling pit initiation rate as a function of environment for Aluminum alloy 7075-T651. *Electrochim. Acta* **2012**, *59*, 336–345. [CrossRef]
71. Lide, D.R. *CRC Handbook of Chemistry and Physics*; CRC Press: Boston, MA, USA, 1990.
72. Hughes, A.E.; Trinchi, A.; Chen, F.F.; Yang, Y.S.; Cole, I.S.; Sellaiyan, S.; Carr, J.; Lee, P.D.; Thompson, G.E.; Xiao, T.Q. Revelation of Intertwining Organic and Inorganic Fractal Structures in Polymer Coatings. *Adv. Mater.* **2014**, *26*, 4504–4508. [CrossRef] [PubMed]
73. Sellaiyan, S.; Hughes, A.E.; Smith, S.V.; Uedono, A.; Sullivan, J.; Buckman, S. Leaching properties of chromate-containing epoxy films using radiotracers, PALS and SEM. *Prog. Org Coat.* **2014**, *77*, 257–267. [CrossRef]
74. Hughes, A.E.; Trinchi, A.; Chen, F.F.; Yang, Y.S.; Cole, I.S.; Sellaiyan, S.; Carr, J.; Lee, P.D.; Thompson, G.E.; Xiao, T.Q. The application of multiscale quasi 4D CT to the study of SrCrO4 distributions and the development of porous networks in epoxy-based primer coatings. *Prog. Org Coat.* **2014**, *77*, 1946–1956. [CrossRef]

© 2017 by the authors. Licensee MDPI, Basel, Switzerland. This article is an open access article distributed under the terms and conditions of the Creative Commons Attribution (CC BY) license (http://creativecommons.org/licenses/by/4.0/).

MDPI
St. Alban-Anlage 66
4052 Basel
Switzerland
Tel. +41 61 683 77 34
Fax +41 61 302 89 18
www.mdpi.com

Coatings Editorial Office
E-mail: coatings@mdpi.com
www.mdpi.com/journal/coatings

www.ingramcontent.com/pod-product-compliance
Lightning Source LLC
LaVergne TN
LVHW070607100526
838202LV00012B/586